Working Guide to Pumps and Pumping Stations

T0229164

Working Guide to Pumps and Pumping Stations

Calculations and Simulations

E. Shashi Menon, P.E.

and

Pramila S. Menon, MBA

AMSTERDAM • BOSTON • HEIDELBERG • LONDON • NEW YORK
OXFORD PARIS • SAN DIEGO • SAN FRANCISCO • SINGAPORE
SYDNEY • TOKYO

Gulf Professional Publishing is an imprint of Elsevier

Gulf Professional Publishing is an imprint of Elsevier
30 Corporate Drive, Suite 400, Burlington, MA 01803, USA
Linacre House, Jordan Hill, Oxford OX2 8DP, UK

Notices
Knowledge and best practice in this field are constantly changing. As new research and
experience broaden our understanding, changes in research methods, professional practices,
or medical treatment may become necessary.
Practitioners and researchers must always rely on their own experience and knowledge
in evaluating and using any information, methods, compounds, or experiments described
herein. In using such information or methods they should be mindful of their own safety and
the safety of others, including parties for whom they have a professional responsibility.
To the fullest extent of the law, neither the Publisher nor the authors, contributors, or
editors, assume any liability for any injury and/or damage to persons or property as a matter
of products liability, negligence or otherwise, or from any use or operation of any methods,
products, instructions, or ideas contained in the material herein.

Library of Congress Cataloging-in-Publication Data

British Library Cataloguing-in-Publication Data
A catalogue record for this book is available from the British Library.

ISBN: 978-1-85617-828-0

For information on all Gulf Professional Publishing
publications visit our Web site at www.elsevierdirect.com

Printed and bound by CPI Group (UK) Ltd, Croydon, CR0 4YY
Transferred to digital print 2013

Working together to grow
libraries in developing countries

www.elsevier.com | www.bookaid.org | www.sabre.org

ELSEVIER   BOOK AID International   Sabre Foundation

Dedicated to our parents

Contents

Preface

This book is about the application of pumps and pumping stations used in pipelines transporting liquids. It is designed to be a working guide for engineers and technicians dealing with centrifugal pumps in the water, petroleum, oil, chemical, and process industries. The reader will be introduced to the basic theory of pumps and how pumps are applied to practical situations using examples of simulations, without extensive mathematical analysis. In most cases, the theory is explained and followed by solved example problems in both U.S. Customary System (English) and SI (metric) units. Additional practice problems are provided in each chapter as further exercise.

The book consists of nine chapters and nine appendices. The first chapter introduces the reader to the various types of pumps used in the industry, the properties of liquids, performance curves, and the Bernoulli's equation. The next chapter discusses the performance of centrifugal pumps in more detail, including variation with impeller speed and diameter. The concept of specific speed is introduced and power calculations explained. Chapter 3 reviews the effect of liquid specific gravity and viscosity on pump performance and how the Hydraulic Institute Method can be used to correct the pump performance for high viscosity liquids. The temperature rise of a liquid when it is pumped and pump operation with the discharge valve closed are discussed.

Chapter 4 introduces the various methods of calculating pressure loss due to friction in piping systems. The Darcy equation, friction factor, the Moody diagram, and the use of the two popular equations for pressure drop (Hazen-Williams and Colebrook-White) are reviewed, and several examples illustrating the method of calculation are solved. Minor losses in valves and fittings, and equivalent lengths of pipes in series and parallel, are explained using example problems. Chapter 5 introduces pipe system head curves and their development, as well as how they are used with the pump head curves to define the operating point for a specific pump and pipeline combination. Chapter 6 explains Affinity Laws for centrifugal pumps and how the pump performance is affected by variation in pump impeller diameter or speed. The method of determining the impeller size or speed required to achieve a specific operating point is explained using examples.

Chapter 7 introduces the concept of net positive suction head (NPSH) and its importance in preventing cavitation in centrifugal pumps. Using examples, the method of calculating the NPSH *available* in a piping system versus the NPSH *required* for a specific pump is illustrated. Chapter 8 covers several applications and economics of centrifugal pumps and pipeline systems. Pumps in series and parallel configuration as well as several case studies for increasing pipeline throughput using additional pumps and pipe loops are discussed. Economic analysis, considering the capital cost, operating and maintenance costs, and rate of return on investment for the most cost effective option are discussed. Finally, Chapter 9 reviews pump simulation using the popular commercial software package PUMPCALC (www.systek.us).

The appendices consist of nine sections and include a list of all formulas presented in the various chapters, unit and conversion factors, the properties of water and other common liquids, the properties of circular pipes, a table of head loss in water pipes, the Darcy friction factor, and the least squares method (LSM) for fitting pump curve data.

Those who purchase this book may also download additional code from the publisher's website for quickly simulating some of the pump calculations described in this book.

We would like to thank Dan Allyn and Barry Bubar, both professional engineers with extensive experience in the oil and gas industry, for reviewing the initial outline of the book and making valuable suggestions for its improvement.

We would also like to take this opportunity to thank Kenneth McCombs, Senior Acquisitions Editor of Elsevier Publishing, for suggesting the subject matter and format for the book. We enjoyed working with him as well as with Andre Cuello, Senior Project Manager, and Daisy Sosa from MPS Content Services.

E. Shashi Menon, P.E.
Pramila S. Menon, MBA
Lake Havasu City, Arizona

Author Biography

E. Shashi Menon, P.E.

E. Shashi Menon, P.E. is Vice President of SYSTEK Technologies, Inc. in Lake Havasu City, Arizona, USA.

He has worked in the Oil and Gas and manufacturing industry for over 37 years. He held positions of design engineer, project engineer, engineering manager and chief engineer with major oil and gas companies in the USA. He has authored three technical books and co-authored over a dozen engineering software applications.

Pramila S. Menon, MBA

Pramila S. Menon, MBA is the President of SYSTEK Technologies, Inc. in Lake Havasu City, Arizona, USA.

She has worked in the Oil and Gas and Financial industry over the last 31 years. Her experience includes systems modeling, feasibility studies, cost analysis and project evaluation.

Introduction

The function of a pump is to increase the pressure of a liquid for the purpose of transporting the liquid from one point to another point through a piping system or for use in a process environment. In most cases, the pressure is created by the conversion of the kinetic energy of the liquid into pressure energy. Pressure is measured in lb/in^2 (psi) in the U.S. Customary System (USCS) of units and in kPa or bar in the Systeme International (SI) system of units. Other units for pressure will be discussed in the subsequent sections of this book. Considering the transportation of a liquid in a pipeline, the pressure generated by a pump at the origin A of the pipeline must be sufficient to overcome the frictional resistance between the liquid and the interior of the pipe along the entire length of the pipe to the terminus B. In addition,

© 2010 E. Shashi Menon. Published by Elsevier Inc. All right reserved.
DOI: 10.1016/B978-1-85617-828-0.00001-9

the pressure must also be sufficient to overcome any elevation difference between A and B. Finally, there must be residual pressure in the liquid as it reaches terminus B if it is to perform some useful function at the end.

If the elevation of B is lower than that of A, there is an elevation advantage where the pump is located that will result in a reduction in the pressure that must be generated by the pump. Conversely, if the elevation of B is higher than that of A, the pump has to work harder to produce the additional pressure to overcome the elevation difference.

Types of Pumps

Several different types of pumps are used in the liquid pumping industry. The most common today is the centrifugal pump, which will be covered in detail in this book. Other types of pumps include reciprocating and rotary pumps. These are called positive displacement (PD) pumps because in each pumping cycle or rotation, the pump delivers a fixed volume of liquid that depends on the geometry of the pump and the rotational or reciprocating speed. In PD pumps, the volume of liquid pumped is independent of the pressure generated. These pumps are able to generate very high pressure compared to centrifugal pumps. Therefore, safety devices such as a rupture disk or a pressure relief valve (PRV) must be installed on the discharge of the PD pumps to protect the piping and equipment subject to the pump pressure.

Centrifugal pumps are capable of providing a wide range of flow rate over a certain pressure range. Hence, the pressure generated by a centrifugal pump depends on the flow rate of the pump. Due to the variation in flow versus pressure, centrifugal pumps are more flexible and more commonly used in process and pipeline applications. They are used in pumping both light and moderately heavy (viscous) liquids. Many applications involving very heavy, viscous liquids, however, may require the use of PD pumps, such as rotary screw or gear pumps, due to their higher efficiency.

Rotary pumps, such as gear pumps and screw pumps, shown in Figure 1.1, are generally used in applications where high-viscosity liquids are pumped. As mentioned before, these PD pumps are able to develop high pressures at fixed flow rates that depend on their design, geometry, and rotational speed.

The operating and maintenance costs of centrifugal pumps are lower compared to PD pumps. In general, PD pumps have better efficiency compared to centrifugal pumps. The recent trend in the industry has been to more often use centrifugal pumps, except for some special high-viscosity and metering applications, where PD pumps are used instead. Since the water pipeline, chemical, petroleum, and oil industries use

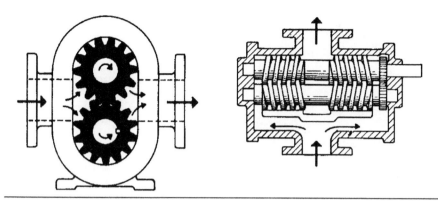

Figure 1.1 Gear pump and screw pump.

mostly centrifugal pumps for their pumping systems, our analysis throughout this book will be geared toward centrifugal pumps.

Centrifugal pumps may be classified into the following three main categories:

■ Radial flow pumps

■ Axial flow pumps

■ Mixed flow pumps

Radial flow pumps develop pressure by moving the pumped liquid radially with respect to the pump shaft. They are used for low flow and high head applications. Axial flow or propeller pumps develop pressure due to the axial movement of the pumped liquid and are used for high flow and low head applications. The mixed flow pumps are a combination of the radial and axial types, and they fall between these two types. The specific speed of a pump, discussed in Chapter 2, is used to classify the type of centrifugal pumps. Radial flow pumps have low specific speeds (less than 2000), while axial flow pumps have high specific speeds (greater than 8000). Mixed flow pumps fall in between.

Figure 1.2 shows a typical centrifugal pump, which can be classified as a volute-type or a diffuser-type pump. The single-volute centrifugal pump in Figure 1.3 converts the velocity head due to the rotational speed of the impeller into static pressure as the liquid is hurled off the rotating impeller into the discharge pipe. Double volute pumps work similar to the single volute type, but they minimize shaft bending due to balanced radial shaft loads. In diffuser-type pumps, stationary guide vanes surround the impeller. These vanes cause gradually expanding passageways for the liquid flow, resulting in gradually changing the direction of the flow and converting the velocity head to pressure head. On the right in Figure 1.3 is a cutaway view of a centrifugal pump coupled to an electric motor driver.

Figure 1.2   Centrifugal pump.

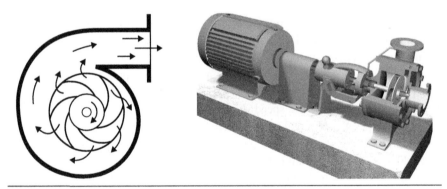

Figure 1.3   Typical cross section of a centrifugal pump.

Liquid Properties

Liquid properties affect the performance of a pump. In this section, some of the important and basic physical properties of liquids that will have a direct bearing on pump performance are reviewed. The three most important liquid properties when dealing with centrifugal pumps are specific gravity, viscosity, and vapor pressure.

SPECIFIC GRAVITY

Specific gravity is a relative measure of the density of a liquid compared to water at the same temperature and pressure. The density is a measure of how heavy a liquid is compared to its volume, or the space it occupies. Thus, we may refer to density in terms of *mass per unit volume*. A related term is *specific weight,* or weight density, which is simply the *weight per unit volume*. If the mass of a certain volume of a liquid is known, dividing the mass by its volume gives the density. You can also obtain the weight density by taking the weight of a certain amount of liquid and dividing it by the volume.

Generally, we tend to use the terms *mass* and *weight* interchangeably. Thus, we talk about a 10-pound mass or a 10-pound weight. Because of this, we can use just the term *density* instead of mass density or weight density. Strictly speaking, mass is a scalar quantity, while weight is a vector quantity that depends on the gravitational force at the location where the measurements are made.

The term *volume* refers to the space occupied by a body. Liquid contained in a tank takes the shape of the container. Thus, a cylindrical tank or a drum full of water has a volume equal to that of the container. In USCS units, volume is stated in cubic feet (ft^3), gallons (gal), or cubic inches (in^3). In SI units, volume may be stated in liters (L) or cubic meters (m^3). The U.S. gallon is equal to $231\,in^3$, whereas the imperial gallon is slightly larger—equal to 1.2 U.S. gallons. Unless specified otherwise, in this book, a *gallon* means a U.S. gallon, denoted as *gal*. In the USCS system of units, mass is stated in pounds (lb), and in SI units, mass is expressed in kilograms (kg). Therefore, the density has the units of lb/ft^3 or lb/gal in USCS units and kg/m^3 or kg/L in SI units.

Since we use pounds (lb) for mass and weight in the USCS units, we will denote the density as lb/ft^3 or lb/gal, depending on whether the volume is measured in ft^3 or gallons. In relation to the SI system, the U.S. gallon can be converted to liters as follows:

$$1\text{ U.S. gallon} = 3.785\text{ L}$$

Here are some other conversions for volume:

$$1\text{ imperial gallon} = 1.2\text{ U.S. gallons} = 1.2\text{ gal}$$

$$1\text{ U.S. gallon} = 231\text{ in}^3 = 0.1337\text{ ft}^3$$

$$1\text{ ft}^3 = 7.4805\text{ gal}$$

In the SI system of units, volume is measured in cubic meters (m^3) or liters (L).
   The density of a liquid in USCS units is

$$\text{Density,}\ \rho = \text{mass/volume} = lb/ft^3$$

In SI units:

$$\text{Density, } \rho = \text{mass/volume} = \text{kg/m}^3$$

The specific gravity (Sg) of a liquid is defined as the ratio of the density of the liquid to the density of water at the same temperature and pressure:

$$Sg = \text{Density of liquid/density of water} \qquad (1.1)$$

Both densities are measured at the same temperature and pressure. Being a ratio of similar entities, specific gravity is dimensionless—in other words, it has no units. The specific gravity of liquids decreases as the temperature increases, and vice versa.

In the petroleum industry, the term *API gravity* (also written as °API) is used in addition to specific gravity. The API gravity, which is always referred to at 60°F, is based on a scale of numbers where water is assigned an API gravity of 10. Products that are lighter than water have an API gravity larger than 10. For example, the API gravity of diesel (Sg = 0.85 at 60°F) is 34.97 °API.

Thus, the API gravity has an inverse relationship with the specific gravity, as follows:

$$\text{Specific gravity } Sg = 141.5/(131.5 + API) \qquad (1.2a)$$

$$API = 141.5/Sg - 131.5 \qquad (1.2b)$$

EXAMPLE 1.1 USCS UNITS

Water weighs 62.4 lb/ft^3, and diesel fuel has a density of 53.04 lb/ft^3 at room temperature and pressure. The relative density or specific gravity of diesel can be calculated using Equation (1.1):

$$Sg = 53.04/62.4 = 0.85$$

To calculate in SI units, for example, suppose a petroleum product weighs 810 kg/m^3, and water at the same temperature and pressure has a density of 995 kg/m^3. Using Equation (1.1), the specific gravity of the product can be found as follows:

$$Sg = 810/995 = 0.8141$$

EXAMPLE 1.2 USCS UNITS

A 55-gal drum containing a petroleum product weighs 339.5 lb after deducting the weight of the drum. What is the density of the liquid and its specific gravity, given that water weighs 62.4 lb/ft^3?

Solution

$$\text{Density of liquid} = \text{mass/volume} = 339.5/55 = 6.1727 \text{ lb/gal}$$

Since 1 ft^3 = 7.4805 gal, we can also express the density as

$$\text{Density} = 6.1727 \times 7.4805 = 46.175 \text{ lb/ft}^3$$

The specific gravity is therefore

$$\text{Sg} = 46.175/62.4 = 0.740$$

VISCOSITY

The viscosity of a liquid is a measure of the liquid's resistance to the flow. Low-viscosity liquids, such as water or gasoline, flow easily in a pipe, compared to heavy, viscous liquids like heavy crude oil, molasses, or asphalt. Therefore, we say that asphalt has a higher viscosity than water. Viscosity may be referred to as dynamic viscosity or kinematic viscosity. Dynamic viscosity, also known as the absolute viscosity, is represented by the Greek letter μ and has the units of poise (P) or centipoise (cP). Kinematic viscosity is represented by the Greek letter ν and has the units of stokes (St) or centistokes (cSt). Both of these units are metric units that are commonly used in both the SI units and the USCS units. Other units of viscosity (visc) are summarized later in this chapter.

Water has an approximate viscosity of 1.0 cP or 1.0 cSt at 60°F, and diesel fuel has a viscosity of approximately 5.0 cSt at 60°F. Like specific gravity, the viscosity of a liquid also decreases as the temperature of the liquid increases, and vice versa.

The viscosity of a liquid can be defined by Newton's equation that states the relationship between the shear stress in the flowing liquid and the velocity gradient of the flow. The constant of proportionality is the dynamic viscosity μ. The velocity gradient occurs because in pipe flow, the velocity of the liquid varies radially at any cross section of the pipe. Imagine a liquid flowing through a transparent pipeline. The liquid molecules adjacent to the pipe wall are at rest or have zero velocity. The liquid molecules that are farthest from the pipe wall—namely, at the center of the pipe—are moving at the maximum velocity. Thus, a velocity variation or a velocity

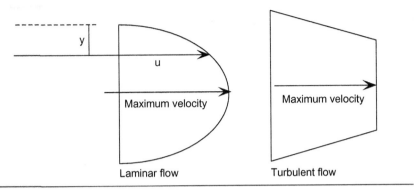

Figure 1.4 Variation of liquid velocity.

gradient exists at any cross section of the pipe. The variation of the velocity at any pipe cross section is represented by the curves in Figure 1.4.

The velocity at a distance of y from the pipe wall is represented by u. The maximum value of u occurs at the centerline of the pipe, where y = the radius of the pipe. The shape of the velocity curve depends on how fast the liquid flows through the pipe and on the type of flow (laminar or turbulent). The velocity variation would approach the shape of a parabola for laminar flow and approximate a trapezoidal shape for turbulent flow. Laminar flow occurs in high-viscosity liquids or at low flow rates. Turbulent flow occurs at higher flow rates and with low-viscosity liquids. The greatest velocity occurs at the centerline of the pipe and can be denoted by u_{max}. The liquid velocity therefore varies from zero at the pipe wall to a maximum of u_{max} at the centerline of the pipe. Measuring the distance y from the pipe wall to a point on the velocity profile, where the velocity is u, we can define the velocity gradient as the rate of change of velocity with the radial distance, or du/dy.

Newton's law states that the shear stress τ between the adjacent layers of the liquid in motion is related to the velocity gradient du/dy as follows:

$$\tau = \mu \, du/dy \tag{1.3}$$

The constant of proportionality is the absolute (dynamic) viscosity of the liquid μ.

The absolute viscosity μ and the kinematic viscosity ν are related by the density of the liquid ρ as follows:

$$\nu = \mu/\rho \tag{1.4}$$

If we choose the units of μ in cP and the units of ν in cSt, the two viscosities are related by the specific gravity Sg as follows:

$$\nu = \mu/Sg \tag{1.5}$$

This simple relationship is due to the convenience of the metric units.

Other units of absolute viscosity μ and the kinematic viscosity ν and some conversions between the units are as follows:

Absolute or dynamic viscosity (μ)

USCS units: lb/ft-s

SI units: poise (P), centipoises (cP) or kg/m-s

Kinematic viscosity (ν)

USCS units: ft^2/s

SI units: stokes (St), centistokes (cSt) or m^2/s

Conversions

$1\,\text{lb-s/ft}^2 = 47.88\,\text{N-s/m}^2 = 478.8\,\text{poise} = 4.788 \times 10^4\,\text{cP}$

$1\,\text{ft}^2/\text{s} = 929\,\text{St} = 9.29 \times 10^4\,\text{cSt}$

$1\,\text{N-s/m}^2 = 10\,\text{poise} = 1000\,\text{cP}$

$1\,\text{m}^2/\text{s} = 1 \times 10^4\,\text{St} = 1 \times 10^6\,\text{cSt}$

In the petroleum industry, two additional sets of viscosity units are employed: Saybolt Seconds Universal (SSU) and Saybolt Seconds Furol (SSF), which are generally used with high-viscosity crude oils or fuel oils. Both are related to the kinematic viscosity, but they do not actually measure the physical property of viscosity. Instead, SSU and SSF represent the time it takes for a fixed volume (usually 60 mL) of the viscous liquid to flow through a specified orifice size at a given temperature. For example, the viscosity of a heavy crude oil at 70°F may be stated as 350 SSU. This means that a 60 mL sample of the crude oil at 70°F in the laboratory took 350 seconds to flow through the specified orifice. SSF is similarly based on the time it takes for a fixed volume of the viscous product to flow through a fixed orifice size at a particular temperature. Both SSU and SSF can be converted to kinematic viscosity in cSt using Equations (1.6) through (1.9)

$$\nu = 0.226 \times \text{SSU} - 195/\text{SSU for } 32 \leq \text{SSU} \leq 100 \qquad (1.6)$$

$$\nu = 0.220 \times \text{SSU} - 135/\text{SSU for SSU} > 100 \qquad (1.7)$$

$$\nu = 2.24 \times \text{SSF} - 184/\text{SSF for } 25 \leq \text{SSF} \leq 40 \qquad (1.8)$$

$$\nu = 2.16 \times \text{SSF} - 60/\text{SSF for SSF} > 40 \qquad (1.9)$$

where ν is the viscosity in centistokes at a particular temperature. Generally, the SSU value is approximately five times the value in cSt. It can be seen from these equations that converting from SSU and SSF to viscosity in cSt is quite straightforward. However, the reverse process is a bit more involved, since a quadratic equation in the unknown quantity SSU or SSF must be solved to determine the value of the viscosity from the given value in cSt. The following example illustrates this method.

EXAMPLE 1.3 USCS UNITS

Diesel fuel has a kinematic viscosity of 5.0 cSt at 60°F, and its specific gravity is 0.85 at 60°F. The dynamic viscosity of diesel can be calculated as follows:

$$\text{Kinematic viscosity, cSt} = \text{dynamic viscosity, cP/Sg}$$

$$5.0 = \mu/0.85$$

Solving for the dynamic viscosity μ, we get

$$\mu = 0.85 \times 5.0 = 4.25 \text{ cP}$$

EXAMPLE 1.4 USCS UNITS

A heavy crude oil at 70°F is reported to have a viscosity of 350 SSU. Calculate its kinematic viscosity in cSt.

Solution

Since the viscosity is greater than 100 SSU, we use the Equation (1.7) for converting to centistokes, as follows:

$$\nu = 0.220 \times 350 - 135/350 = 76.61 \text{ cSt}$$

EXAMPLE 1.5 SI UNITS

A sample of a viscous crude oil was found to have a viscosity of 56 cSt at 15°C. Calculate the equivalent viscosity in SSU.

Solution

Since the SSU value is roughly five times the cSt value, we expect the result to be close to $5 \times 56 = 280$ SSU. Since this is greater than 100 SSU, we use Equation (1.7) to solve for SSU for $\nu = 56$ cSt, as follows:

$$56 = 0.220 \times \text{SSU} - 135/\text{SSU}$$

Rearranging the equation, we get

$$0.220(SSU)^2 - 56(SSU) - 135 = 0$$

Solving this quadratic equation for SSU, we get

$$SSU = (56 + \sqrt{(56^2 + 4 \times 0.22 \times 135)})/(2 \times 0.22) = 256.93$$

We have ignored the second negative root of the quadratic because the viscosity cannot be negative.

VAPOR PRESSURE

The vapor pressure of a liquid is an important property when dealing with centrifugal pumps. Vapor pressure is defined as the pressure of the liquid at a certain temperature when the liquid and its vapor are in equilibrium. Thus, we can say that the boiling point of a liquid is the temperature at which its vapor pressure equals the atmospheric pressure. Generally, the vapor pressure of a liquid is measured at 100°F in the laboratory and is called the Reid Vapor Pressure (RVP).

When you know the RVP of the liquid, the corresponding vapor pressure at any other temperature can be determined using standard charts. The importance of vapor pressure of a liquid will be discussed later in the section when the available suction pressure is calculated. The vapor pressure of water at 60°F is 0.256 psia, and in SI units, the vapor pressure of water at 40°C is 7.38 kPa (abs). Vapor pressure is usually stated in absolute pressure units of psia or kPa (abs). The vapor pressure of a liquid increases as the liquid temperature increases.

SPECIFIC HEAT

The specific heat of a liquid (Cp) is defined as the heat required to raise the temperature of a unit mass of liquid by one degree. It is a function of temperature and pressure. For most liquids that are incompressible, such as water or gasoline, specific heat depends only on the temperature and is found to increase as the temperature increases.

In USCS units, specific heat is expressed in Btu/lb/°F, and in SI units, it is stated in kJ/kg/°C. Water has a specific heat of 1 Btu/lb/°F (4.186 kJ/kg/°C), whereas petroleum products have specific heats ranging between 0.4 and 0.5 Btu/lb/°F (1.67 and 2.09 kJ/kg/°C).

PRESSURE AND HEAD OF A LIQUID

Pressure at any point in a liquid is a function of the depth of that point below the free surface of the liquid. For example, consider a storage tank containing a liquid with the free surface 20 ft above the tank bottom. The pressure in the liquid halfway down the tank is one-half the pressure at the bottom of the tank. The pressure at the surface of the liquid will be designated as zero reference pressure or atmospheric pressure. According to Pascal's Law, all of the points in the liquid that are at the same depth below the free surface will have the same amount of pressure.

The pressure in a liquid is measured using a pressure gauge, and thus the reading is called the gauge pressure. If a pressure gauge is attached to the bottom of a tank that contains 20 ft of liquid, the pressure indicated by the gauge will be approximately 8.66 lb/in^2 gauge (psig). This calculation will be explained shortly.

Since the surface of the liquid in the tank is subject to the local atmospheric pressure (approximately 14.7 psi at sea level), the actual pressure at the bottom of the tank is 8.66 + 14.73 = 23.39 lb/in^2 absolute (psia) (Figure 1.5). Thus, the absolute pressure (psia) is obtained by adding the gauge pressure (psig) to the local atmospheric pressure:

$$P_{abs} = P_{gauge} + P_{atm} \qquad (1.10)$$

In this book, gauge pressure (psig) is implied unless it is explicitly specified as absolute pressure (psia).

In USCS units, pressure is stated in lb/in^2 (psi) or lb/ft^2 (psf). In SI units, pressure is expressed as kilopascal (kPa), megapascal (MPa), or bar. See Appendix B for conversion factors between various units of pressure.

The atmospheric pressure at a location depends on its geographic elevation above some reference point, such as mean sea level (MSL). The atmospheric pressure decreases with altitude, ranging from 101.3 kPa at MSL to 66.1 kPa at an altitude of 3500 m. In USCS units, the atmospheric pressure is approximately 14.7 psi at MSL and drops to 10.1 psi at an altitude of 10,000 ft.

The pressure at a point in a liquid will increase with the depth in a linear manner. For example, pressure P at a depth h below the free surface is

$$P = h \times Sg/2.31 \qquad (1.11)$$

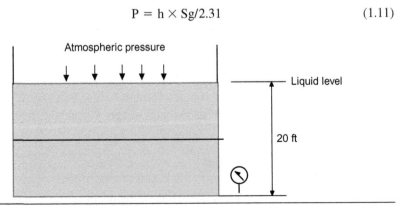

Figure 1.5   Pressure of liquid in a tank.

where

P: Pressure, psig

Sg: Specific gravity of liquid, dimensionless

h: Depth below free surface of liquid, ft

The corresponding equation for SI units is as follows:

$$P = h \times Sg/0.102 \tag{1.12}$$

where

P: Pressure, kPa

Sg: Specific gravity of liquid, dimensionless

h: Depth below free surface of liquid, m

From Equation (1.11) it is clear that the pressure in a liquid is directly proportional to the depth h. The latter is also referred to as the head of a liquid. Thus, a pressure of 1000 psi is equivalent to a certain head of liquid. The head, which is measured in ft (or m in SI), depends on the liquid specific gravity. Considering water (Sg = 1.00), the head equivalent of a pressure of 1000 psi is calculated from Equation (1.11) as

$$2.31 \times 1000/1.0 = 2310 \text{ ft}$$

Thus, the pressure of 1000 psi is said to be equivalent to a head of 2310 ft of water. If the liquid were gasoline (Sg = 0.736), the corresponding head for the same 1000 psi pressure is

$$2.31 \times 1000/0.736 = 3139 \text{ ft}$$

We can see that as the liquid specific gravity decreases, the head equivalent for a given pressure increases. Alternatively, for a heavier liquid such as brine (Sg = 1.25), the corresponding head for 1000 psi pressure is

$$2.31 \times 1000/1.25 = 1848 \text{ ft}$$

Figure 1.6 illustrates the effect of the liquid specific gravity on the head, in ft of liquid, for a given pressure in psig.

Using Equation (1.12) to illustrate an example in the SI units, a pressure of 7000 kPa (70 bar) is equivalent to a head of

$$7000 \times 0.102/1.0 = 714 \text{ m of water}$$

In terms of gasoline, this is equal to

$$7000 \times 0.102/0.736 = 970.11 \text{ m of gasoline}$$

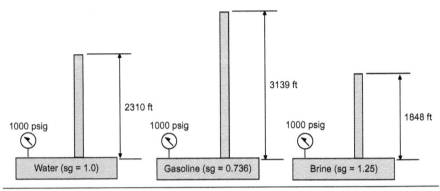

Figure 1.6 Pressure versus head for different liquids.

Energy of a Flowing Liquid and Bernoulli's Equation

We will first discuss the flow of a liquid through a pipe to introduce the concepts of the three energy components of a flowing liquid and Bernoulli's equation. In Figure 1.7, a liquid flows through a pipe from point A to point B at a uniform rate of flow Q. At point A, a unit mass of the liquid has three components of energy:

1. Pressure energy due to the liquid pressure
2. Kinetic energy due to the velocity of flow
3. Potential energy due to the elevation of the liquid above some datum

The principle of Conservation of Energy states that energy is neither created nor destroyed but is simply converted from one form of energy to another. Bernoulli's equation, which is just another form of the same principle, states that the total energy of the liquid as it flows through a pipe at any point is a constant. Thus, in Figure 1.7, the total energy of the liquid at point A is equal to the total energy of the liquid at point B, assuming no energy is lost in friction or heat and that there is no addition of energy to the liquid between these two points. These are the three components of energy at A:

1. Pressure energy due to the liquid flow or pressure: P_A/γ
2. Kinetic energy due to the flow velocity: $V_A^2/2g$
3. Potential energy due to the elevation: Z_A

The term P_A/γ is the *pressure head*, $V_A^2/2g$ is the *velocity head*, and Z_A is the *elevation head*. The term γ is the specific weight of the liquid, which is assumed to be the same at A and B, since liquids are generally considered to be incompressible. The term γ will change with the temperature, just as the liquid density changes with temperature. In USCS units, γ is stated in lb/ft³, and in SI units, it is expressed in

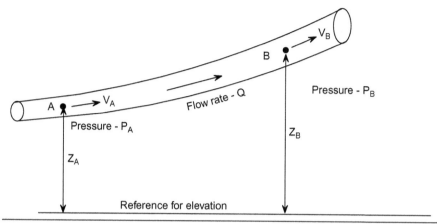

Figure 1.7   Energy of a flowing liquid.

kN/m^3. For example, the specific weight of water at 20°C is $9.79\,kN/m^3$. The term g in the kinetic energy is the acceleration due to gravity. It is a constant equal to $32.2\,ft/s^2$ in USCS units and $9.81\,m/s^2$ in SI units. According to Bernoulli's equation, if we neglect frictional losses in the pipe, in steady flow, the sum of the three components of energy is a constant. Therefore,

$$P_A/\gamma + V_A^2/2g + Z_A = P_B/\gamma + V_B^2/2g + Z_B \qquad (1.13)$$

where P_A, V_A, and Z_A are the pressure, velocity, and elevation head, respectively, at point A. Similarly, subscript B refers to the point B in Figure 1.7.

Let us examine each term of the equation. In USCS units, the pressure head term has the units of $lb/ft^2/lb/ft^3 = ft$, the velocity head has the units of $(ft/s)^2/ft/s^2 = ft$, and the elevation head has the units of ft. Thus, all of the terms of the Bernoulli's equation have the units of head: ft.

In SI units, the pressure head term has the units of $kN/m^2/kN/m^3 = m$. Similarly, the velocity head has the units of $(m/s)^2/m/s^2 = m$, and the elevation head has the units of m. Thus, all of the terms of the Bernoulli's equation have the units of head, m, and each term in the equation represents the energy in units of liquid head in ft in USCS units or m in SI units.

In reality, we must take into account the energy lost due to friction in pipe flow. Therefore, Equation (1.13) has to be modified as follows:

$$P_A/\gamma + V_A^2/2g + Z_A - h_f = P_B/\gamma + V_B^2/2g + Z_B \qquad (1.14)$$

where h_f is the frictional pressure drop, or head loss in the pipe due to liquid flow, between A and B. Similarly, if we add energy to the liquid at some point between A and B, such as using a pump, we must add that to the left-hand side of Equation

(1.14). Considering the pump head as h_p we can rewrite Bernoulli's equation, taking into account frictional head loss and the pump head, as follows:

$$P_A/\gamma + V_A^2/2g + Z_A - h_f + h_p = P_B/\gamma + V_B^2/2g + Z_B \qquad (1.15)$$

EXAMPLE 1.6 USCS UNITS

A water pipeline like the one in Figure 1.7 has a uniform inside diameter of 15.5 inches, and the points A and B are 4500 ft apart. Point A is at an elevation of 120 ft, and point B is at an elevation of 350 ft. The flow rate is uniform, and the velocity of flow is 5.4 ft/s.

(a) If the pressure at A is 400 psi, and the frictional head loss between A and B is 32.7 ft, calculate the pressure at point B.

(b) If a pump that is midway between A and B adds 220 ft of head, what is the pressure at B, assuming the data given in (a)?

Solution

(a) Using Bernoulli's equation (1.14), we get

$$400 \times 144/(62.4) + 120 - 32.7 \text{ ft} = P_B \times 144/(62.4) + 350$$

Solving for pressure at B,

$$P_B \times 144/(62.4) = 923.08 + 120 - 32.7 - 350 = 660.38$$

$$P_B = 660.38 \times 62.4/144 = 286.16 \text{ psi}$$

(b) With the pump adding 220 ft of head, the pressure at B will be increased by that amount, as follows:

$$P_B = 286.16 \text{ psi} + 220 \times 1/2.31 = 381.4 \text{ psi}$$

Pump Head and Capacity

Because of the direct relationship between pressure and head, pump vendors standardized water as the pumped liquid and refer to the pump pressure in terms of head of water. Thus, a centrifugal pump is said to develop a certain amount of head in ft of water for a given capacity or flow rate in gal/min (gpm) at a certain pump efficiency. For example, in USCS units the pump vendor may indicate that a certain model of pump has the capability of producing a head of 1400 ft and an efficiency of 78% at a capacity of 500 gal/min. Of course, this is what can be accomplished with water as the liquid pumped. As long as the liquid is not much more viscous than water, the same head and capacity will be realized when pumping diesel or gasoline. However, if a heavier, more viscous (greater than 10 cSt) product such as a heavy crude oil is to be pumped, the head, capacity, and efficiency will be different. Generally, the higher the viscosity of the pumped liquid, the lower the head generated and the lower the efficiency at a given capacity. This is called viscosity corrected performance and will be addressed in more detail in Chapter 3.

As an example of a pump specification in SI units, the pump vendor may state that a certain model of pump produces a head of 420 m and an efficiency of 75% at a capacity of 108 m³/h.

EXAMPLE 1.7 USCS UNITS

For a pumping application, a pressure of 350 psi is required at a flow rate of 49 ft³/min of diesel (Sg = 0.85). How would the pump be specified for this application in pump vendor terminology?

Solution

Convert 350 psi to ft of liquid head:

$$H = 350 \times 2.31/0.85 = 952 \text{ ft, rounding up}$$

The capacity of the pump in gal/min is

$$Q = 49 \times 1728/231 = 367 \text{ gal/min}$$

Thus, a suitable pump for this application should generate 952 ft of head at 367 gal/min at an efficiency of 80% to 85%, depending on the pump model.

EXAMPLE 1.8 SI UNITS

A pumping application requires a flow rate of 40 L/s of gasoline (Sg = 0.736) at a pressure of 18 bar. How would you specify this pump?

Solution

Convert 18 bar to head in meters by transposing Equation (1.12):

$$H = 18 \times 100 \times 0.102/0.736 = 250 \text{ m, rounding up}$$

The capacity of the pump in m³/h is

$$Q = 40 \times 3600/1000 = 144 \text{ m}^3/\text{h}$$

Thus, a suitable pump for this application should generate 250 m head at 144 m³/h at an efficiency of 80% to 85%, depending on the pump model.

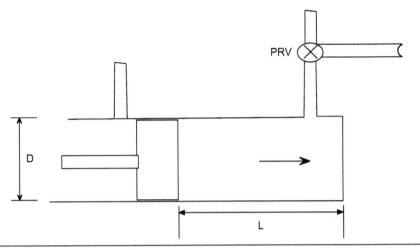

Figure 1.8 Positive displacement piston pump.

As we indicated at the beginning of this chapter, the primary function of a pump is to develop pressure in a liquid to perform some useful work in a process or to move the liquid through a pipeline from a source, such as a storage tank, to a destination, such as another storage tank. Now we will examine how this is done in a PD pump compared to a centrifugal pump.

In a PD pump, such as a reciprocating piston type pump, during each stroke of the piston, a fixed volume of liquid is delivered from the suction piping to the discharge piping at the required pressure, depending on the application. The exact volume of liquid delivered depends on the diameter of the cylinder, the stroke of the piston, and the speed of the pump, as shown in Figure 1.8.

A graphic plot of the pressure versus the volume of liquid pumped by a PD pump is simply a vertical line, as shown in Figure 1.9, indicating that the PD pump can provide a fixed flow rate at any pressure, limited only by the structural strength of the pump and the attached piping system. Due to pump clearances, leaks, and the liquid viscosity, there is a slight drop in volume as the pressure increases, called slip, as shown in Figure 1.9.

It is clear that theoretically this PD pump could develop as high a pressure as required for the application at the fixed volume flow rate Q. The upper limit of the pressure generated will depend on the strength of the pipe to which the liquid is delivered. Thus, a pressure relief valve (PRV) is attached to the discharge side of the PD pump, as shown in Figure 1.8. It is clear that in such an installation, we can get only a fixed flow rate unless the geometry of the pump is changed by replacing it with a larger-sized pump or increasing its speed.

A centrifugal pump, on the other hand, is known to provide a flexible range of flow rates and pressures. A typical pressure (head) versus flow rate (capacity) curve

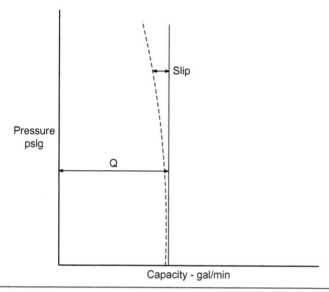

Figure 1.9   Capacity versus pressure for a PD pump.

for a centrifugal pump is shown in Figure 1.10. It can be seen that the useful range of capacities for this centrifugal pump is 500 to 1500 gpm, with heads ranging from 1200 to 800 ft.

To generate pressure, liquid enters the suction side of a centrifugal pump at a certain initial pressure (suction pressure). The energy of the liquid corresponding to this pressure combined with the kinetic energy of the liquid due to suction velocity and the potential energy due to the position of the suction piping represent the total energy of the liquid when it enters the pump. As the liquid flows through the pump, it is accelerated by the rotation of the pump impeller due to centrifugal force, and the kinetic energy of the liquid is increased. Finally, as the liquid exits, the kinetic energy is converted to pressure energy as it exits the pump volute into the discharge piping.

It is clear that the flow rate through the pump and the pressure generated are both functions of the rotational speed of the pump impeller and its diameter, since it is the impeller that adds to the kinetic energy of the liquid. Thus, the capacity of the centrifugal pump may be increased or decreased by changing the pump speed within certain limits. The upper limit of speed will be dictated by the stresses generated on the pump supports and the limitation of the pump drivers such as the electric motor, turbine, or engine. Also, the pump capacity and head developed by the pump will increase with the diameter of the impeller. Obviously, due to the space limitation within the casing of the pump, there is a practical limit to the maximum size of the impeller. Centrifugal pump performance will be discussed in more detail in subsequent chapters.

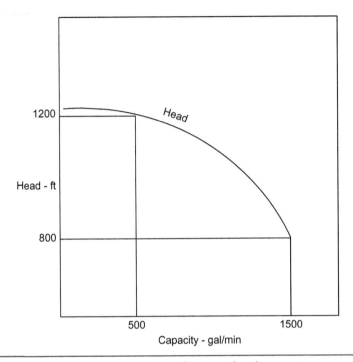

Figure 1.10 Flow rate versus pressure for centrifugal pump.

Summary

This chapter provided an introduction to the different types of pumps used in the industry. The differences between PD pumps and centrifugal pumps were explained, along with the advantages of the centrifugal pump. The process of creating pressure in a PD pump compared to a centrifugal pump was explained, as were several important liquid properties that affect pump performance. The various units of viscosity of a liquid and the concept of vapor pressure and its importance were reviewed. The terms *pressure* and *head* of a liquid were explained, and several example problems were solved to further illustrate the various concepts introduced in the chapter. In the next chapter the performance of centrifugal pumps will be examined and analyzed in more detail.

Problems

1.1 The specific gravity of diesel is 0.85 at 60°F. Water has a density of 62.4 lb/ft^3. Calculate the weight of a 55-gal drum of diesel.

1.2 A heavy crude oil was found to have specific gravity of 0.89 at 60°F. What is the equivalent API gravity?

1.3 The kinematic viscosity of a petroleum product is 15.2 cSt at 15°C. If its API gravity is 42.0, calculate its dynamic viscosity at 15°C.

1.4 The pressure gauge connected to the bottom of a tank containing water shows 22 psig. The atmospheric pressure at the location is 14.5 psi. What is the actual absolute pressure at the bottom of the tank? What is the water level in the tank?

1.5 In the previous problem, if the liquid is gasoline (Sg = 0.736 at 16°C), what pressure reading in bar will be indicated for a 16 m liquid level in the tank?

1.6 The suction and discharge pressure gauges on a centrifugal pump pumping water (Sg = 1.0) indicates 1.72 bar and 21.5 bar, respectively. What is the differential head produced by the pump in meters of water?

Chapter 2

Pump Performance

In Chapter 1, we briefly discussed performance of PD pumps and centrifugal pumps. In this chapter we will review in more detail the performance of centrifugal pumps and their variation with pump speed and impeller size. The concept of pump-specific speed will be discussed, and how pumps are selected for a specific application will be explained using examples.

The performance of a centrifugal pump is characterized by graphic plots showing the head (pressure) developed by the pump versus capacity (flow rate), pump efficiency versus capacity, pump brake horsepower (BHP) versus capacity, and NPSH versus capacity, as shown in Figure 2.1.

In the USCS units, the pressure developed by a pump at any capacity (Q) in ft of liquid head (H) is plotted on the vertical axis, and the flow rate or capacity (Q) in

DOI: 10.1016/B978-1-85617-828-0.00002-0

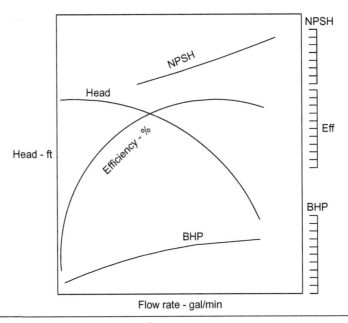

Figure 2.1 Centrifugal pump performance curves.

gal/min is plotted along the horizontal axis. In SI units, the head is usually plotted in meters (m) and capacity is plotted in m³/h.

The terms *capacity* and *flow rate* are used interchangeably, with pump vendors preferring the term *capacity*. Since pumps are used in applications involving a multitude of different liquids, to standardize, the manufacturer's pump curves are always based on water (Sg = 1.0 and Visc = 1.0 cSt) as the pumped liquid. The performance curves are then modified as necessary for a specific liquid pumped, depending on its specific gravity and viscosity.

Since the pressure developed by a centrifugal pump is plotted in ft of liquid head, in USCS units (or m in SI units), it will have the same head characteristic as long as the liquid is not very viscous (visc < 10 cSt). Therefore, if the water performance for a particular model pump is stated as 2300 ft of head at a capacity of 1200 gpm, the same pump will develop 2300 ft of head when pumping diesel (visc = 5.0 cSt) or light crude oil (visc = 8.9 cSt), since the viscosities are less than 10 cSt. If the viscosity of the liquid is higher than 10 cSt, the head curve will have to be modified from the water performance curve to handle the higher viscosity liquid, as will be discussed in Chapter 3.

As we saw in Chapter 1, when the capacity Q varies, the head H developed by the centrifugal pump also varies, as indicated in Figure 2.2. Typically, centrifugal pumps have a drooping H-Q curve, indicating that at zero flow, maximum head is generated. This is known as the shutoff head. With the increase in flow, the head

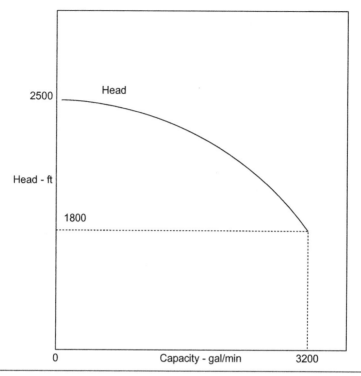

Figure 2.2   Head-capacity curve.

decreases until the minimum head is generated at the maximum capacity, as shown in Figure 2.2.

It can be seen from Figure 2.2 that the shutoff head is 2500 ft (at Q = 0), and at maximum capacity Q = 3200 gal/min, the head is 1800 ft. The shape of the H-Q curve is approximately a parabola. Therefore, mathematically the H-Q curve can be represented by the second-degree polynomial equation

$$H = a_0 + a_1 Q + a_2 Q^2 \tag{2.1}$$

where a_0, a_1, and a_2 are constants for the pump that depend on the pump geometry, impeller diameter, and speed, respectively.

It is important to realize that the head versus capacity (H-Q) curve shown in Figure 2.2 is based on a particular pump impeller diameter running at a specific rotational speed, or revolutions per minute (RPM). A larger impeller diameter would produce a higher head and capacity. Similarly, a smaller diameter impeller will generate a lower H-Q curve, as shown in Figure 2.3, for a fixed pump speed of 3560 RPM.

Also in Figure 2.3, we assume that the pump is driven by an electric motor at some fixed speed, such as 3560 RPM. If we increase the pump speed, the H-Q

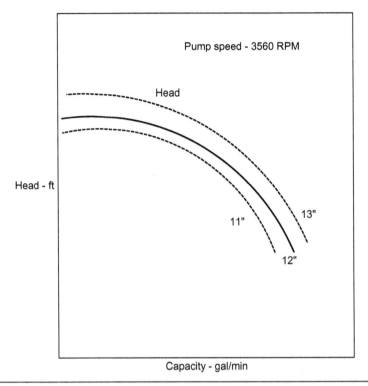

Figure 2.3 H-Q curve at different impeller diameters.

curve will be higher. If we lower the pump speed, the H-Q curve will be lower. This is similar to the variation with impeller diameter, as indicated in Figure 2.4.

To recap, Figure 2.3 shows the pump head generated for various capacities (solid curve) for a 12-inch impeller running at 3560 RPM. Replacing this impeller with a smaller 11-inch-diameter impeller results in the lower dashed H-Q curve shown. Similarly, replacing the current 12-inch impeller with a 13-inch impeller results in the H-Q curve shown by the upper dashed curve. It must be noted that we are comparing the performance of the same pump at various impeller diameters, running at a fixed speed of 3560 RPM.

Let us now examine the effect of varying the pump speed while keeping the impeller diameter fixed. Figure 2.4 shows a pump head curve for a 10-inch-diameter impeller diameter running at a speed of 3560 RPM, represented by the solid curve. By keeping the impeller diameter fixed at 10 in. but slowing the pump speed to 3000 RPM, we get the lower dashed H-Q curve. Likewise, increasing the speed to 4000 RPM results in the upper dashed H-Q curve.

We will next review the variation of pump efficiency with capacity, or the E-Q curve. The efficiency of a pump indicates how effectively the energy supplied to the pump by the drive motor is utilized in generating the pressure head in the liquid.

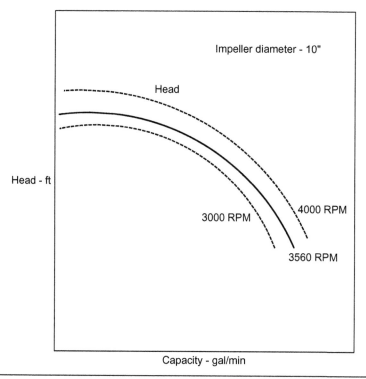

Figure 2.4   H-Q curve at different impeller speeds.

The pump efficiency is a function of the pump internal geometry and impeller diameter. It does not vary significantly with the impeller speed.

A typical efficiency curve for a centrifugal pump starts off at zero efficiency for zero capacity (Q = 0) and reaches the maximum efficiency (known as the best efficiency) at some capacity and then drops off as the capacity continues to increase up to the maximum pump capacity, as shown in Figure 2.5. The capacity at which the highest efficiency is attained is called the Best Efficiency Point (BEP) and is usually designated on the H-Q curve as BEP. Thus, the BEP, designated by a small triangle, represents the best operating point on the pump H-Q curve that results in the highest pump efficiency, as shown in Figure 2.5. Generally, in most centrifugal pump applications, we try to operate the pump at a capacity close to and slightly to the left of the BEP. In the petroleum industry, there is a preferred operating range (POR) for centrifugal pumps that is 40% to 110% of the BEP flow rate.

The E versus Q curve can also be represented by a parabolic equation, similar to that of the H-Q curve:

$$E = b_0 + b_1 Q + b_2 Q^2 \tag{2.2}$$

where b_0, b_1, and b_2 are constants for the pump.

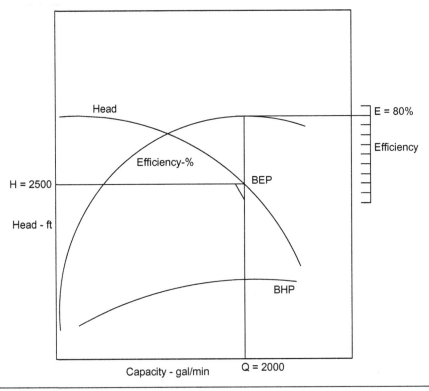

Figure 2.5 Efficiency – capacity curve.

Equations (2.1) and (2.2) are used to approximate a typical centrifugal pump head and efficiency curves for simulation on an Excel spreadsheet or in a computer program. When we are given a pump vendor's performance curve data, we can develop the two equations for H and E as functions of Q for further analysis of pump performance.

Manufacturers' Pump Performance Curves

Most pump manufacturers publish a family of performance curves for a particular model and size of pump, as shown in Figure 2.6. These curves show a series of H-Q curves for a range of pump impeller diameters, along with a family of constant efficiency curves, known as the iso-efficiency curves, and constant Power curves. These are all drawn for a particular size and model (example size 6 × 8 × 15-DVM) of a centrifugal pump operating at a fixed RPM. Sometimes these curves include the constant NPSH curves as well.

In addition to the performance curves shown in Figure 2.6, pump manufacturers also publish composite rating charts for a family of centrifugal pumps, as shown in

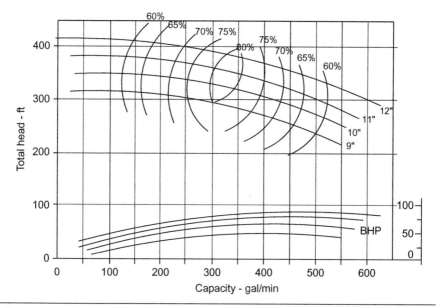

Figure 2.6   Typical pump manufacturer's performance curves.

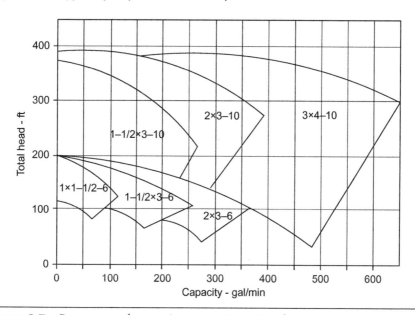

Figure 2.7   Pump manufacturer's composite rating charts.

Figure 2.7. These composite rating charts show the range of capacities and heads that a line of pumps of a particular design can handle for different sizes, such as 1 × 1–1/2–6 to 3 × 4–10. From these charts, we can pick a pump size to handle the capacity and head range we anticipate for our application. For example, if an application requires a pump that can develop 200 ft of head at a capacity of 250 gpm, there will be more than one size of pump that can meet the requirement. However, because the performance ranges overlap, there will be only one particular pump size that will be the optimum for the desired duty. In all cases, the manufacturer must be consulted to make the final selection of the pump, particularly since they can provide more data on potential future modifications that may be needed for system expansions.

EXAMPLE 2.1 USCS UNITS

The head H versus capacity Q of a centrifugal pump is represented by the following data:

Q (gal/min)	550	1100	1650
H (ft)	2420	2178	1694

Develop an equation for the H-Q curve and calculate the pump shutoff head.

Solution

Using Equation (2.1), we can substitute the three pairs of H-Q values and obtain the following equation in the unknowns a_0, a_1, and a_2:

$$2420 = a_0 + 550a_1 + (550)^2 a_2$$
$$2178 = a_0 + 1100a_1 + (1100)^2 a_2$$
$$1694 = a_0 + 1650a_1 + (1650)^2 a_2$$

Solving the three simultaneous equations, we get

$$a_0 = 2420 \quad a_1 = 0.22 \quad a_2 = -0.0004$$

Therefore, the equation for the H-Q curve is

$$H = a_0 + a_1 Q + a_2 Q^2 = 2420 + 0.22Q - 0.0004Q^2$$

The shutoff head H_0 can be calculated from Equation (2.1) by setting $Q = 0$:

$$H_0 = a_0 = 2420$$

EXAMPLE 2.2 SI UNITS

The following data was taken from a pump manufacturer's pump curve:

Q - L/min	600	750	900
H - m	216	200	172
E - %	80.0	82.0	79.5

Develop an equation for the E-Q curve.

Solution

Using Equation (2.2), we can substitute the three pairs of E-Q values and obtain the following three equations in b_0, b_1, and b_2:

$$80 = b_0 + 600b_1 + (600)^2 b_2$$
$$82 = b_0 + 750b_1 + (750)^2 b_2$$
$$79.5 = b_0 + 900b_1 + (900)^2 b_2$$

Solving the three simultaneous equations, we get

$$b_0 = 27.025 \qquad b_1 = 0.1483 \qquad b_2 = -1.0 \times 10^{-4}$$

Therefore, the equation for the E-Q curve is

$$E = 27.025 + 0.1483Q - Q^2/10^4$$

In Examples 2.1 and 2.2, we developed an equation for H versus Q and E versus Q from the three sets of pump curve data. Since we had to determine the values of the three constants (a_0, a_1, and a_2) for the H-Q curve, the given set of three pairs of data was just about adequate for the three simultaneous equations. If we had selected four or more sets of pump curve data, we would have more equations than the number of unknowns. This would require a different approach to solving for the constants a_0, a_1, and a_2, as illustrated in Example 2.3.

EXAMPLE 2.3 USCS UNITS

The following sets of pump curve data are from a pump manufacturer's catalog:

Q gal/min	1000	2000	3000	4000	5000	6000	7000
H ft	1445	1430	1410	1380	1340	1240	1100
E %	32.5	51.5	63	69	72.5	73	71.5

Determine the best fit curve for the H-Q and E-Q data.

Solution

Here we have seven sets of Q, H, and E data for determining the values of the constants in the H-Q and E-Q Equations (2.1) and (2.2).

Obviously, if we substituted the seven sets of H-Q data in Equation (2.1), we will have seven simultaneous equations in three unknowns: a_0, a_1, and a_2. Thus, we have more equations to solve simultaneously than the number of variables. In such a situation we need to determine the best-fit curve for the data given. Using the least squares method (LSM), we can determine the optimum values of the constants a_0, a_1, and a_2 for the H-Q equation and the constants b_0, b_1, and b_2 for the E-Q equation. The least squares method is based on minimizing the square of the errors between the actual value of H and the theoretical, calculated value of H, using Equation (2.1). For a detailed discussion of the least squares method and fitting data to curves, refer to a book on numerical analysis or statistics.

Using LSM, which is described in summary form in Appendix I, we calculate the constants as follows:

$$a_0 = 1397.9 \quad a_1 = 0.0465 \quad a_2 = -0.00001$$

Therefore, the equation for the H-Q curve is

$$H = 1397.9 + 0.0465Q - 0.00001Q^2$$

Similarly, for the E-Q curve, using the least squares method, we calculate the constants as follows:

$$b_0 = 14.43 \quad b_1 = 0.0215 \quad b_2 = -0.0000019$$

Therefore, the equation for the E-Q curve is

$$E = 14.43 + 0.0215Q - 0.0000019Q^2$$

To test the accuracy of these equations, we can substitute the values of Q into the equations and compare the calculated values H and E with the given data. For example, at $Q = 1000$, we get

$$H = 1397.9 + 0.0465 \times 1000 - 0.00001 \times (1000)^2 = 1434.4 \text{ ft}$$

Compare this with a given value of 1445. The difference is less than 1%.

Similarly, for $Q = 1000$, we calculate E as follows:

$$E = 14.43 + 0.0215 \times 1000 - 0.0000019 \times (1000)^2 = 34.03\%$$

This compares with the given value of 32.5%, with the error being approximately 4.7%. Therefore, the LSM approach gives a fairly good approximation for the pump head H and efficiency E as a function of the capacity Q, using the given pump curve data.

Power Required by a Pump: Hydraulic and Brake Horsepower

Since the pump develops a head H at a capacity Q, the power required by the pump is proportional to the product of H and Q. If the efficiency is assumed to be 100%, the power required is referred to as hydraulic horsepower (HHP) in USCS units. In SI units, it is called the hydraulic power in kW. When the efficiency of the pump is included, we get a more accurate estimate of the actual power required by the pump, also known as brake horsepower (BHP). In SI units, it is called the brake power and is stated in kW. The conversion between the two units is as follows:

$$1 \text{ HP} = 0.746 \text{ kW}$$

$$\text{and } 1 \text{ kW} = 1.34 \text{ HP}$$

In USCS units, considering water as the liquid pumped, the HHP is calculated as follows:

$$\text{HHP} = Q \times H \times Sg/3960 \qquad (2.3)$$

where

Q: capacity, gal/min
H: head, ft
Sg: specific gravity of liquid pumped, dimensionless

In SI units, the hydraulic power required in kW is as follows:

$$\text{Hydraulic power (kW)} = Q \times H \times Sg/(367.46) \qquad (2.4)$$

where

Q: capacity, m^3/h
H: head, m
Sg: specific gravity of liquid pumped, dimensionless

This is the theoretical power required considering the pump efficiency as 100%. The brake horsepower (BHP) or the brake power (SI units) is calculated by including the efficiency in the denominator of Equations (2.3) and (2.4), respectively, as follows:

$$\text{BHP} = Q \times H \times Sg/(3960 \times E) \qquad (2.5)$$

where

Q: capacity, gal/min
H: head, ft
Sg: specific gravity of liquid pumped, dimensionless
E: pump efficiency (decimal value, less than 1.0)

In SI units, the pump brake power required in kW is as follows:

$$\text{Power} = Q \times H \times Sg/(367.46 \times E) \tag{2.6}$$

where

Q: capacity, m^3/h
H: head, m
Sg: specific gravity of liquid pumped, dimensionless
E: pump efficiency (decimal value, less than 1.0)

When water is the pumped liquid, the BHP calculated is called the water horse-power (WHP). If the pumped liquid is something other than water, the BHP can be calculated by multiplying the WHP by the liquid specific gravity.

It must be noted that the BHP calculated is the power required by the pump that is transferred as pressure to the liquid pumped. The driver of the pump, which may be an electric motor, a turbine, or a diesel engine, must provide enough power to the pump. Based on the driver efficiency, its power will be more than the pump power calculated. An electric drive motor may have an efficiency of 95%, which can be used to determine the electric power input to the motor, knowing the BHP of the pump. For example, if the BHP calculated for a pump is 245 HP, the electric motor that drives the pump will require a power input of 245/0.95 at 95% motor efficiency. This works out to 258 HP or 258 × 0.746 = 193 kW. The electric energy consumption can easily be determined if you know the kW input to the motor.

Similar to the H-Q and the E-Q curves, we can plot the BHP versus Q curve for a pump using the equations for power introduced earlier in this chapter. For each value of Q from the pump curve data, we can obtain the head H and efficiency E and calculate the BHP from Equation (2.5). Thus, a list of BHP values can be cal-culated and tabulated for each of the capacity values Q. The resulting plot of BHP versus Q can be superimposed on the H-Q and E-Q curve as shown in Figure 2.8.

Generally, the BHP curve is a rising curve, meaning that the BHP increases with an increase in capacity and usually tapers off at the maximum pump capacity. However, in some cases, depending on the shape of the head and the efficiency curves, the peak power requirement may not be at the highest capacity but at

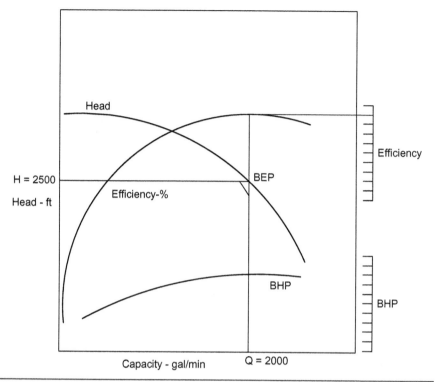

Figure 2.8   BHP versus capacity curve.

a lower value. Therefore, in selecting the driver motor size, the pump power requirement over the entire capacity range must be determined, rather than that at the maximum capacity.

Referring to the H-Q curve based on Equation (2.1) and noting that the BHP is the product of Q and H, we can readily see that the BHP can be represented by a third-degree polynomial in Q, as follows:

$$BHP = Q(a_0 + a_1Q + a_2Q^2)Sg/(K \times E) \qquad (2.7)$$

where K is a constant. Since the efficiency E is also a quadratic function of Q, from Equation (2.7), we can rewrite the BHP equation as follows:

$$BHP = Q(a_0 + a_1Q + a_2Q^2)Sg/[K(b_0 + b_1Q + b_2Q^2)]$$

By suitable mathematical manipulation, we can reduce this equation to a simpler equation as follows:

$$BHP = (c_0 + c_1Q + c_2Q^2)Sg \qquad (2.8)$$

where c_0, c_1, and c_2 are constants to be determined from the pump curve data.

EXAMPLE 2.4 USCS UNITS

A centrifugal pump has the following performance data at the BEP:

Q = 2300 gal/min
H = 2100 ft
E = 78%

Calculate the HHP and BHP at the BEP, considering gasoline (sg = 0.740).

Solution

Using Equation (2.3) for HHP, we calculate

$$HHP = 2300 \times 2100 \times 0.74 /(3960) = 902.58$$

For BHP, we use Equation (2.5) and calculate

$$BHP = 2300 \times 2100 \times 0.74/(3960 \times 0.78) = 1157.15$$

The WHP for this pump at the BEP can be calculated based on water in place of gasoline as follows:

$$WHP = 2300 \times 2100 \times 1.0/(3960 \times 0.78) = 1563.71$$

EXAMPLE 2.5 SI UNITS

A centrifugal pump has the following values at the BEP:

Q = 520 m³/hr
H = 610 m
E = 82%

Calculate the power required at the BEP when pumping crude oil with specific gravity of 0.89.

Solution

Using Equation (2.6), we calculate the power required in kW as follows:

$$Power = 520 \times 610 \times 0.89/(367.46 \times 0.82) = 936.91\,kW$$

EXAMPLE 2.6 USCS UNITS

Using the following pump curve data from a pump manufacturer's catalog, develop the BHP curve for water. What size electric motor driver should be selected? If the pump operated continuously for 24 hours a day for 350 days a year at the BEP, calculate the annual energy cost at $0.10/kWh.

Q gpm	1000	2000	3000	4000	5000	6000	7000
H ft	1445	1430	1410	1380	1340	1240	1100
E %	32.5	51.5	63	69	72.5	73	71.5

Solution

Using Equation (2.5), we calculate the BHP_1 through BHP_7 for the preceding seven sets of data. For Q = 1000, H = 1445, and E = 32.5%, the BHP for water (Sg = 1.0) is as follows:

$$BHP_1 = 1000 \times 1445 \times 1.0/(3960 \times 0.325) = 1123 \text{ (rounded up)}$$

Repeating the preceding calculations for the remaining six sets of H, Q, and E values, we can develop the following data set for the BHP curve (note that all BHP values have been rounded off to the nearest whole number):

Q gpm	1000	2000	3000	4000	5000	6000	7000
BHP	1123	1403	1696	2020	2334	2574	2720

The BHP versus capacity Q curve can now be plotted from the preceding data.

It can be seen from the BHP values that the highest pump power required is 2720 BHP at the maximum capacity. If the drive motor has an efficiency of 95%, the motor HP required is

$$\text{Motor power} = 2720/0.95 = 2863\,HP = 2863 \times 0.746 = 2136\,kW$$

Examining the given data, the BEP is at Q = 6000, H = 1240, and E = 73%. The BHP calculated at BEP from the preceding table = 2574. The annual energy consumption = 2574 × 0.746 × 24 × 350 = 16,129,714 kWh. The annual energy cost at $0.10 per kWh = $0.1 × 16,129,714 = $1.61 million.

NPSH versus Pump Capacity

The fourth curve that is part of the family of pump curves is the one that shows the variation of the NPSH versus pump capacity. NPSH stands for the net positive

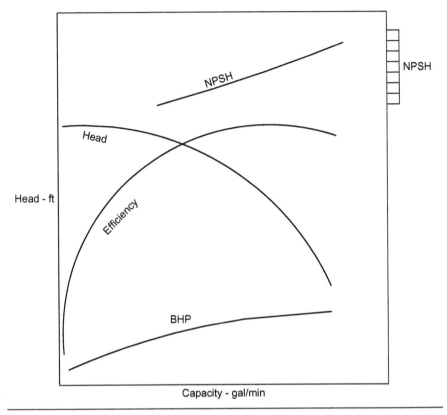

Figure 2.9 NPSH versus capacity curve.

suction head required for a pump at a particular capacity. It is a measure of the minimum suction head required at the suction of the pump impeller above the liquid vapor
pressure. NPSH is a very important parameter when pumping high-vapor-pressure
liquids. The concept of NPSH, its impact on pump performance, and examples of
how to calculate the available NPSH versus the minimum required for a specific
pump are all discussed in more detail in Chapter 7. The shape of the NPSH versus
Q curve is a gradually rising curve as shown in Figure 2.9.

Pump Driver and Power Required

The power calculations show how much power is required to drive the pump at
each capacity value. The maximum power usually occurs at the maximum capacity
of the pump. In the previous example, the highest BHP of 2720 is required at the
maximum capacity of Q = 7000 gal/min. The drive motor for this pump will have

to provide at least this much power. In fact, if we take into account the efficiency of the electric motor (usually around 95%), the installed motor HP must be at least

$$2720/0.95 = 2863 \text{ HP}$$

Usually, for safety, a 10% allowance is included in the calculation of the installed HP required. This means that, in this example, motor power is increased by 10% to $1.1 \times 2863 = 3150$ HP. The nearest standard size electric motor for this application is 3200 HP.

When an electric motor is used to drive a centrifugal pump, the electric motor may have a name plate service factor that ranges from 1.10 to 1.15. This service factor is an indication that the motor may safely handle an overload of 10% to 15% in an emergency situation compared to the name plate rating of the motor. Thus, a 1000-HP electric motor with a 1.15 service factor can provide a maximum power of $1.15 \times 1000 = 1150$ HP under emergency conditions without the motor windings burning out.

Centrifugal pumps are usually driven by constant speed motors or some form of a variable speed driver. The latter may be an engine, a turbine, or a variable frequency electric motor. The majority of pump drives use constant speed electric motors, which are the least expensive of the drives. The constant speed motors usually drive the pump at speeds of approximately 1800 RPM or 3600 RPM. The actual speed depends on the type of electric motor (synchronous or induction) and is a function of the electrical frequency and the number of poles in the motor. The frequency of the electric current used is either 60 Hz (U.S.) or 50 Hz (U.K. and many other countries).

The synchronous speed of an electric motor can be calculated knowing the electrical frequency f and the number of poles p in the motor as follows:

$$Ns = 120 \times f/p \qquad (2.9)$$

Considering a 4-pole motor and a frequency of 60 Hz, the synchronous speed of the motor is

$$Ns = 120 \times 60/4 = 1800 \text{ RPM}$$

Similarly, for a 2-pole motor, the synchronous speed is

$$Ns = 120 \times 60/2 = 3600 \text{ RPM}$$

For 50 Hz frequency, these become

$$Ns = 120 \times 50/4 = 1500 \text{ RPM}$$

for a 4-pole motor, and

$$Ns = 120 \times 50/2 = 3000 \text{ RPM}$$

for a 2-pole motor.

In induction motors, the actual motor speed is slightly below the calculated synchronous speed, based on the frequency and number of poles. The difference between the two speeds is referred to as the slip. Therefore, a typical 4-pole induction motor will run at a constant speed of about 1780 RPM, whereas the 2-pole motor will run around 3560 RPM.

Variable speed drive (VSD) is a general term that refers to a pump driver that has a range of speeds—for example, from 2000 RPM to 4000 RPM. These include electric motors with variable frequency drive (VFD), gas turbine drives, and engine drives. There are also less expensive VSDs that use fluid couplings, currently in use with pipeline applications. The maximum and minimum pump speeds permissible will also be specified by the pump manufacturer for a particular model pump and impeller size. This is because the pump performance depends on the impeller speed. Also at higher impeller speeds, increased centrifugal forces cause higher stresses on the components of the pump.

Multistage Pumps

Due to limitations of the head that can be produced with a particular impeller diameter and rotational speed, additional head can be achieved only by using multiple pumps or increasing the number of stages in a pump. A single-stage pump may develop 400 ft of head at 1000 gal/min. Suppose we require a head of 1200 ft at a flow rate of 1000 gal/min. We can use a three-stage pump that develops 400 ft of head per stage. Multistage pumps are used when a single stage will not be sufficient to handle the head requirement. Alternatively, an existing multistage pump can be destaged to reduce the head developed for a particular application.

For example, let us say that an application requires a pump to provide 500 ft of head at 800 gal/min. A three-stage pump is available that develops 750 ft of head at 800 gal/min. If we destage this pump from three stages to two stages, we will be able to reduce the head developed at the same flow rate to two-thirds of 750 ft, or 500 ft. Destaging affects the pump efficiency, and therefore the pump vendor must be contacted for verification. Also, if the liquid pumped is viscous, the performance will be different from that of the water performance. In some instances it might not be possible to increase the number of stages within a pump. In such cases, the head required can be produced only by utilizing more than one pump in series. Multiple pumps in series and parallel configurations are discussed in more detail in Chapter 8.

Specific Speed

An important parameter called the specific speed is used to compare different pump models for different applications. The specific speed is a function of the pump impeller speed, head, and capacity at the BEP. The specific speed is calculated using the following formula:

$$N_S = N\,Q^{\frac{1}{2}} / H^{\frac{3}{4}} \qquad (2.10)$$

where

N_S: Specific speed of the pump
N: Impeller speed, RPM
Q: Capacity at BEP, gal/min
H: Head per stage at BEP, ft

In SI units, Equation (2.10) is used for specific speed, except Q will be in m³/h and H in m.

It is clear that there is some inconsistency in the units used in Equation (2.10). For consistent units, we would expect N_S and N to have the same units of speed. Since the term $(Q^{\frac{1}{2}} / H^{\frac{3}{4}})$ is not dimensionless, the units of N_S will not be the same as the speed N. Regardless, this is the way specific speed has been defined and used by the industry.

On examining the equation for specific speed, we can say that the specific speed is the speed at which a geometrically similar pump must be run so that it develops a head of 1 foot at a capacity of 1 gal/min. It must be noted that the Q and H used in Equation (2.10) for specific speed refer to the values at the best efficiency point (BEP) of the pump curve based on the maximum impeller diameter. When calculating the specific speed for a multistage pump, H is the head per stage of pump. Additionally, we infer from the specific speed equation that at high H values the specific speed is low and vice versa. Thus, high-head pumps have low specific speeds and low-head pumps have high specific speeds. The specific speeds of centrifugal pumps are in the range of 500 to 20,000 and depend on the design of the pump. Radial flow pumps have the lowest specific speed, while axial flow pumps have the highest specific speed. The mixed flow pumps have specific speeds between the two types. Typical values of specific speeds are listed in Table 2.1.

Another version of the specific speed parameter for centrifugal pumps that uses the NPSH at the BEP instead of the head per stage at BEP is called the suction specific speed. This parameter is calculated as follows:

$$N_{SS} = N\,Q^{\frac{1}{2}} / (NPSH_R)^{\frac{3}{4}} \qquad (2.11)$$

where

N_{SS}: Suction specific speed of the pump
N: Impeller speed, RPM
Q: Capacity at BEP, gal/min
NPSH_R: NPSH required at BEP, ft

Table 2.1   Specific Speeds of Centrifugal Pumps

Pump Type	Specific Speed	Application
Radial Vane	500–1000	Low capacity/high head
Francis—Screw Type	1000–4000	Medium capacity/ medium head
Mixed—Flow Type	4000–7000	Medium to high capacity, low to medium head
Axial—Flow Type	7000–20,000	High capacity/low head

In SI units, the same formula is used, with Q in m³/h and NPSH$_R$ in m. The term *NPSH* used in Equation (2.11) is an important parameter in preventing pump cavitation and will be discussed in detail in Chapter 7.

When dealing with double suction pumps, it must be noted that the value of Q used to calculate the two specific speeds N$_S$ and N$_{SS}$ are not the same. For calculating the specific speed (N$_S$), the value of Q at BEP is used. However, for calculating the suction specific speed, (Q/2) is used for double suction pumps.

EXAMPLE 2.7 USCS UNITS

Consider a 5-stage pump running at 3570 RPM, with the following BEP values: Q = 2000 gal/min and H = 2500 ft. Calculate the specific speed of the pump.

Solution

Using Equation (2.10), we calculate the specific speed as follows:

$$N_S = 3570 \times (2000)^{\frac{1}{2}} / (2500/5)^{\frac{3}{4}} = 1510$$

EXAMPLE 2.8 USCS UNITS

Calculate the specific speed of a 4-stage double suction centrifugal pump, 12 in. diameter impeller that has a speed of 3560 RPM, and develops a head of 2000 ft at a flow rate of 2400 gal/min at the BEP. Also, calculate the suction specific speed if the NPSH required is 25 ft at the BEP.

Solution

Using Equation (2.10), the specific speed is calculated as follows:

$$N_S = N\,Q^{\frac{1}{2}} / H^{\frac{3}{4}}$$

$$= 3560\,(2400)^{\frac{1}{2}} / (2000/4)^{\frac{3}{4}} = 1650$$

From Equation (2.11), the suction specific speed is calculated as follows:

$$N_{SS} = N\,Q^{\frac{1}{2}} / NPSHR^{\frac{3}{4}}$$

$$= 3560\,(2400/2)^{\frac{1}{2}} / (25)^{\frac{3}{4}} = 11,030$$

EXAMPLE 2.9 SI UNITS

Calculate the specific speed of a 4-stage single suction centrifugal pump, 250-mm diameter impeller that has a speed of 2950 RPM and develops a head of 600 m at a flow rate of 540 m³/h at the BEP. Also, calculate the suction specific speed if the NPSH required is 6.6 m at the BEP.

Solution

Using Equation (2.10), the specific speed is calculated as follows:

$$N_S = N\,Q^{\frac{1}{2}} / H^{\frac{3}{4}}$$

$$= 2950\,(540)^{\frac{1}{2}} / (600/4)^{\frac{3}{4}} = 1600$$

From Equation (2.11), the suction specific speed is calculated as follows:

$$N_{SS} = N\,Q^{\frac{1}{2}} / NPSH_R^{\frac{3}{4}}$$

$$= 2950\,(540)^{\frac{1}{2}} / (6.6)^{\frac{3}{4}} = 16,648$$

Summary

In this chapter we introduced the various performance curves for a centrifugal pump such as head, efficiency, power, and NPSH as a function of the flow rate through the pump. The approximations for the H-Q and E-Q curves as a parabola were discussed and illustrated using examples. The variations of the head generated by a centrifugal pump with changes in impeller diameter and impeller speed were reviewed. The difference between hydraulic HP and brake horsepower was explained using examples. The importance of NPSH was explained along with pump power and driver power required. The specific speed concept was introduced and illustrated using examples. In the next chapter, the performance of a centrifugal pump and how it is affected by the liquid viscosity will be discussed in more detail.

Problems

2.1 A centrifugal pump has the following H-Q and E-Q data taken from the pump curve. Determine the coefficient for a second-degree polynomial curve using LSM.

Q gpm	600	1200	2400	4000	4500
H ft	2520	2480	2100	1680	1440
E %	34.5	55.7	79.3	76.0	72.0

Calculate the HHP at the maximum capacity. What is the BHP when pumping water at the BEP?

2.2 A centrifugal pump has the following BEP condition:

$$Q = 480 \text{ m}^3/\text{h} \quad H = 420 \text{ m} \quad E = 81.5\%$$

(a) Calculate the power in kW when pumping water at the BEP.
(b) If this same pump was used to pump gasoline (Sg = 0.74), what power is required at the BEP?

2.3 Calculate the specific speed of a centrifugal pump running at 3560 RPM with the following BEP values:

$$Q = 520 \text{ gpm} \quad H = 280 \text{ ft} \quad E = 79.5\%$$

2.4 A three-stage double suction centrifugal pump, 10 in. impeller diameter operates at a speed of 3570 RPM and develops 2400 ft at the BEP capacity of 3000 gpm. Calculate the specific speed of this pump. What is the suction specific speed if NPSH required is 21 ft?

Liquid Properties versus Pump Performance

In the previous chapter, it was mentioned that the manufacturers' pump performance curves are always referred to in terms of water. In this chapter we will review the performance of a pump when pumping liquids other than water and also explore the effect of high viscosity on pump performance. We will also discuss the temperature rise of a liquid due to pumping and the effect of running a pump for a short time with the discharge valve closed. Finally, knowing the pipe length, flow rate, and pressure loss due to friction, we will explain how a pump is selected for a particular application. First, we will examine how the pump performance varies with the basic properties such as specific gravity and viscosity of the liquid.

Consider a pump head curve that has a shutoff head of 2500 ft and a maximum capacity of 3000 gal/min. Suppose also that the BEP of this pump curve is at

$Q = 2800$, $H = 1800$, and $E = 82.0\%$. The manufacturer's catalog would indicate that for low-viscosity liquids (visc < 10 cSt), the performance would be the same as that of water. In other words, for a product such as gasoline or diesel, the head and efficiency curves for this pump will remain the same. Therefore, this pump will have the same BEP whether it is pumping water ($Sg = 1.0$) or gasoline ($Sg = 0.736$) or diesel ($Sg = 0.85$). In other words, this pump can produce a head of 1800 ft at the BEP when pumping water or diesel ($Sg = 0.85$) at a flow rate of 2800 gpm. However, due to the difference in specific gravity of the two liquids, the actual pressure in psi when pumping water at this flow rate will be

$$\text{Pressure} = 1800 \times 1/2.31 = 779.22 \text{ psi}$$

And when pumping diesel will be

$$\text{Pressure} = 1800 \times 0.85/2.31 = 662.34 \text{ psi}$$

Therefore, even though the pump develops the same head in ft regardless of the liquid pumped, the pressure in psi will be different for different liquids. Similarly, in SI units, the head in m will be the same for all liquids, but the pressure in kPa or bar will be different for different liquids. In general, as long as the viscosity of the liquid is low (less than 10 cSt), the pump efficiency will also be the same regardless of the liquid pumped. Thus, the H-Q curve and the E-Q curve are the same for water, gasoline, or diesel, since the viscosities of these liquids are all less than 10.0 cSt. As the viscosity of the liquid increases, the H-Q curve and the E-Q curve have to be corrected if the viscosity is above 10 cSt, using the Hydraulic Institute method. The BHP curve will depend on the specific gravity of the liquid as indicated in Equation (2.5). For example, when pumping water, the preceding defined pump at BEP requires the following BHP:

$$\text{BHP}_\text{w} = 2800 \times 1800 \times 1.0/(3960 \times 0.82) = 1552$$

When pumping diesel:

$$\text{BHP}_\text{d} = 2800 \times 1800 \times 0.85/(3960 \times 0.82) = 1319$$

And for gasoline:

$$\text{BHP}_\text{g} = 2800 \times 1800 \times 0.736/(3960 \times 0.82) = 1142$$

This is illustrated in Figure 3.1, where the head, efficiency, and BHP curves for water, diesel, and gasoline are shown.

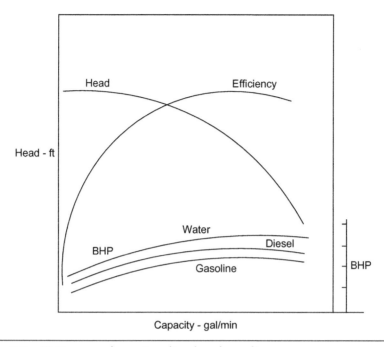

Figure 3.1   BHP curves for water, diesel and gasoline.

When the head is plotted in ft of liquid, the H-Q curve for the three liquids are the same, as is the efficiency E-Q. The BHP curve for water is at the top, for gasoline at the lowest, and for diesel in between. Therefore, for moderately viscous (viscosity $< 10\,cSt$) liquids, the physical properties such as specific gravity and viscosity do not affect the H-Q and E-Q curves compared to those for water. The BHP curves are different because BHP is directly proportional to the specific gravity of the liquid, as indicated by Equation (2.5).

Let us now examine the effect of pumping a high-viscosity liquid. Figure 3.2 shows the head, efficiency, and BHP curves for a typical pump. The solid curves are for water performance, while the dashed curves are the corrected performance curves when pumping a viscous liquid (viscosity $> 10\,cSt$). These corrected curves were generated using the Hydraulic Institute charts for viscosity-corrected performance, discussed in detail in the following pages.

It can be seen from Figure 3.2 that when pumping a viscous liquid, the head versus capacity curve is located slightly below the H-Q curve for water, indicating that the head generated at a certain same capacity is lower for a viscous liquid, compared to the head developed when pumping water. For estimating the viscous performance, a correction factor C_H (<1.0) must be applied to the head values from the water curve. Similarly, the efficiency curve, when pumping the higher viscosity liquid, is also lower than when pumping water. The correction factor for the efficiency curve, which

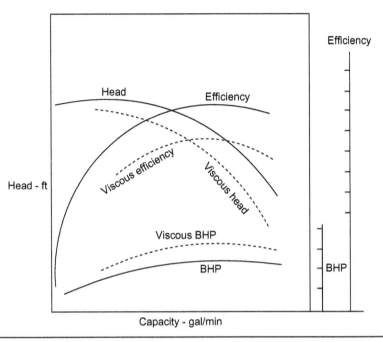

Efficiency

Head

Efficiency

Head - ft

Viscous efficiency

Viscous head

Viscous BHP

BHP

BHP

Capacity - gal/min

Figure 3.2 Pump performance corrected for viscous liquid.

is denoted by C_E, is also a number less than 1.0. Generally, the efficiency correction factor C_E is a smaller value than the head correction factor C_H, indicating that the effect of the high viscosity is to degrade the pump efficiency more than the head.

Actually, there is also a capacity correction factor C_Q that is applied to the capacity values of the water curve, so all three parameters—Q, H, and E—from the water curve need to be reduced by the respective correction factors to determine the viscosity corrected performance of the pump. These correction factors depend on the liquid viscosity and the Q, H, and E values at the BEP for the water curve. They are usually calculated using the Hydraulic Institute method. The Hydraulic Institute handbook outlines the method of derating a water performance curve when pumping a high-viscosity liquid, using a set of charts, as described in the following pages.

Starting with the water performance curve for a pump, using the factors C_Q, C_H, and C_E obtained from the Hydraulic Institute charts, new H-Q and E-Q curves can be generated for the viscous liquid. It should be noted that while the viscous curves for H-Q and E-Q are located below the corresponding curves for water, the viscous BHP curve is located above that of the water curve, as shown in Figure 3.2. Note that the correction factors C_Q and C_E are fixed for the capacity range of the pump curve. However, the correction factor C_H varies slightly over the range of capacity, as will be explained shortly.

In order to estimate the viscosity corrected performance of a pump, we must know the specific gravity and viscosity of the liquid at the pumping temperature along with the H-Q curve and E-Q curve for water performance. Knowing the BEP on the pump curve, three additional capacity values, designated at 60%, 80%, and 120% of BEP capacity, are selected on the head curve. For example, if the BEP point is at $Q = 1000\,\text{gal/min}$, the Q values 600, 800, 1000, and 1200 gal/min are selected. For the selected Q values, the H values and E values are tabulated by reading them off the water performance curve.

If the pump is a multistage unit, the head values are reduced to the head per stage. For each H, Q, and E value, we obtain the correction factors C_Q, C_H, and C_E for the liquid viscosity from the Hydraulic Institute chart. Using the correction factors, a new set of Q, H, and E values are generated by multiplying the corresponding values from the water curve by the correction factors. The process is repeated for the four sets of Q values (60%, 80%, 100%, and 120%). The revised set of Q, H, and E values forms the basis of the viscosity corrected pump curves. An example will illustrate this method.

Commercial software programs are available to estimate the viscosity corrected performance of a pump easily and quickly, without resorting to the Hydraulic Institute

---

EXAMPLE 3.1 USCS UNITS

A 5-stage centrifugal pump has the following data taken from the manufacturer's catalog. Using Hydraulic Institute chart, determine the corrected performance when pumping crude oil with the following properties: Specific gravity = 0.95 and viscosity = 660 cSt at 60°F. The water performance is plotted in Figure 3.3.

Q (gal/min)	H (ft)	E (%)
727	2275	54.2
1091	2264	68.2
1455	2198	76.9
1818	2088	81
2000	2000	82
2182	1896	81.1
2546	1648	76.1
2909	1341	70

Solution

By inspection, the BEP for this pump curve is

$$Q = 2000 \quad H = 2000 \text{ and } E = 82\%$$

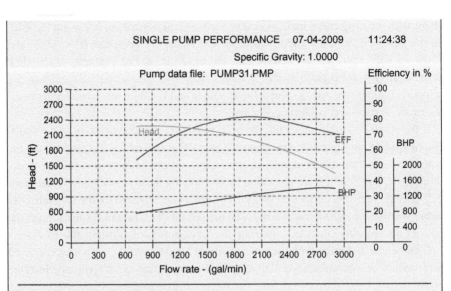

Figure 3.3 Water performance curve.

The head per stage at BEP is $H_{stg} = 2000/5 = 400$ ft.

Next, determine the four sets of Q values at which the H and E values will be read off the water performance curve:

At 60% BEP value:

$$Q_{60} = 0.6 \times 2000 = 1200 \, \text{gal/min}$$

At 80% BEP value:

$$Q_{80} = 0.8 \times 2000 = 1600 \, \text{gal/min}$$

At 100% BEP value:

$$Q_{100} = 2000 \, \text{gal/min}$$

At 120% BEP value:

$$Q_{120} = 1.2 \times 2000 = 2400 \, \text{gal/min}$$

Next, determine from the H-Q and E-Q plots of the water curve the interpolated values of H and E for the four sets of Q values:

$$\text{For } Q_{60} = 1200 \quad H = 2250 \quad E = 71.4$$
$$\text{For } Q_{80} = 1600 \quad H = 2162 \quad E = 78.9$$
$$\text{For } Q_{100} = 2000 \quad H = 2000 \quad E = 82.0$$
$$\text{For } Q_{120} = 2400 \quad H = 1755 \quad E = 78.4$$

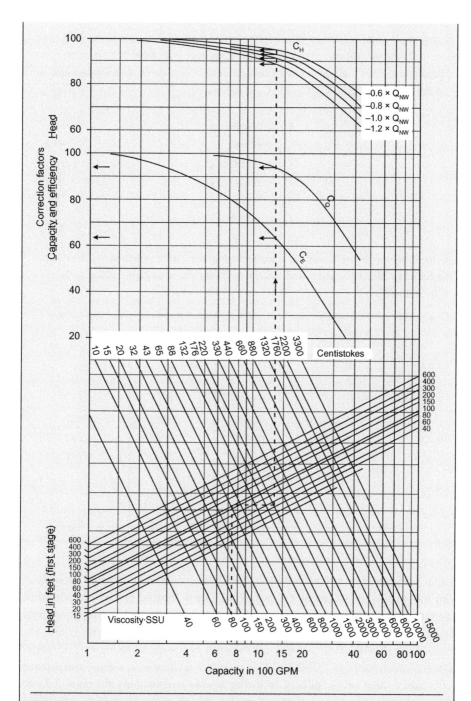

Figure 3.4   Hydraulic institute viscosity correction chart.
"Courtesy of the Hydraulic Institute, Parsippany, NJ, www.Pumps.org"

Using the Hydraulic Institute charts for viscosity correction, we locate the BEP flow rate of Q = 2000, and go vertically to the head per stage line of 400 ft. Next, move horizontally to the left until the viscosity line of 660 cSt, then move vertically upwards to get the correction factors C_Q, C_H, and C_E as follows:

$$C_E = 0.5878$$
$$C_Q = 0.9387$$
$$C_H = 0.9547 \text{ at } Q_{60}$$
$$C_H = 0.9313 \text{ at } Q_{80}$$
$$C_H = 0.9043 \text{ at } Q_{100}$$
$$C_H = 0.8718 \text{ at } Q_{120}$$

Using these correction factors, we compile the following table, which shows the water performance, the correction factors, and the viscosity corrected performance for the 660 cSt liquid.

	$0.6 \times Q_{NW}$	$0.8 \times Q_{NW}$	$1.0 \times Q_{NW}$	$1.2 \times Q_{NW}$
Q_W	1200.0	1600.0	2000.0	2400.0
H_W	2250.0	2162.0	2000.0	1755.0
E_W	71.4	78.9	82.0	78.4
C_Q	0.9387	0.9387	0.9387	0.9387
C_H	0.9547	0.9313	0.9043	0.8718
C_E	0.5878	0.5878	0.5878	0.5878
Q_V	1126.5	1502.0	1877.5	2253.0
H_V	2148.1	2013.4	1808.6	1530.0
E_V	42.0	46.4	48.2	46.1

charts. These programs require inputting the basic pump H-Q and E-Q data, along with the number of stages of the pump and liquid specific gravity and viscosity. The output from the program consists of the viscosity corrected pump performance data and graphic plots of the performance curves with water and viscous liquid. The results using PUMPCALC (www.systek.us) software are shown in Figure 3.5. A brief review of the features of PUMPCALC is included in Chapter 9, where several pump performance cases are simulated, including viscosity correction. In Figure 3.5, the water performance curves are shown, along with the viscous performance curves for a typical centrifugal pump.

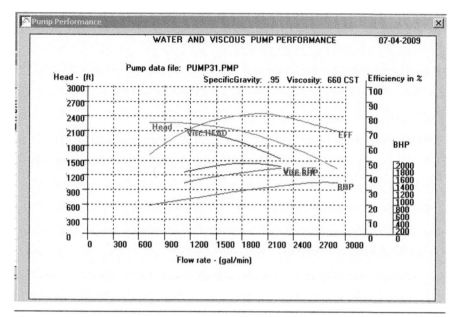

Figure 3.5   Water performance versus viscous performance.

Temperature Rise of Liquid Due to Pump Inefficiency

It is clear that the efficiency of a pump is less than 100%, and therefore a portion of the energy supplied to the pump by the driver is converted to friction. This results in a slight increase in the temperature of the liquid as it goes through the pump. The amount of heating imparted to the liquid can be estimated from the head, efficiency, and specific heat of the liquid, as explained in this equation:

$$\Delta T = H(1/E - 1)/(778 Cp)$$

(3.1)

where

ΔT: temperature rise of liquid from suction to discharge of pump, °F
H: head at the operating point, ft
E: efficiency at the operating point, (decimal value, less than 1.0)
Cp: liquid specific heat, Btu/lb/°F

The efficiency E used in Equation (3.1) must be a decimal value (less than 1.0), not a percentage. It can be seen that the temperature rise is zero when the pump efficiency is 100% (E = 1.0).

In SI units the temperature rise of the pumped liquid is calculated from

$$\Delta T = H(1/E - 1)/(101.94 Cp)$$

(3.2)

where

ΔT: temperature rise of liquid from suction to discharge of pump, °C
H: head at the operating point, m
E: efficiency at the operating point (decimal value, less than 1.0)
Cp: liquid specific heat, kJ/kg/°C

As before, the efficiency E used in Equation (3.2) must be a decimal value (less than 1.0), not a percentage. It can be seen from the equation that the temperature rise is zero when the pump efficiency is 100% (E = 1.0).

EXAMPLE 3.2 USCS UNITS

A centrifugal pump is operating very close to its BEP as follows:

$$Q = 1800 \, \text{gal/min} \qquad H = 2200 \, \text{ft} \qquad E = 78\%$$

The specific heat of the liquid is Cp = 0.45 Btu/lb/°F. Calculate the temperature rise of the liquid due to pumping.

Solution
Using Equation (3.1), we get the temperature rise as

$$\Delta T = 2200(1/0.78 - 1)/(778 \times 0.45) = 1.77°F$$

EXAMPLE 3.3 SI UNITS

Calculate the temperature rise of a liquid due to pumping for a centrifugal pump that is operating at its BEP as follows:

$$Q = 115 \, \text{L/s} \qquad H = 700 \, \text{m} \qquad E = 79\%$$

The specific heat of the liquid is Cp = 1.89 kJ/kg/°C.

Solution
Using Equation (3.2), we get the temperature rise as

$$\Delta T = 700 \, (1/0.79 - 1)/(101.94 \times 1.89) = 0.97°C$$

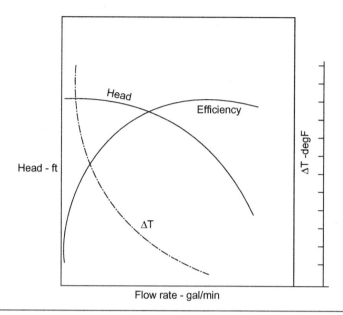

Figure 3.6   Temperature rise of liquid due to pumping.

Examining Equation (3.1), it is clear that the temperature rise when pumping a liquid depends on the operating point on the pump curve—namely, the head H and efficiency E. As the operating point shifts to the left of the BEP, the temperature rise is higher than when the operating point is to the right of the BEP. The variation of ΔT with capacity Q is shown in Figure 3.6.

The temperature rise increases to a very high value as the flow rate through the pump is reduced to shutoff conditions. The effect of operating a pump with a closed discharge valve is discussed next.

Starting Pump against a Closed Discharge Valve

When a centrifugal pump starts up, it is usually started with the discharge valve closed to prevent overload of the drive motor, since the flow rate is practically zero and the power required, and thus the demand, on the electric motor drive is minimal. If the pump runs against a closed valve for too long, there is a danger of overheating the liquid, which in turn could cause vaporization. Vaporized liquid causes damage to the pump impeller, since pumps are designed to pump liquids, not gases. Hence, overheating of the pumped liquid and potential vaporization should be avoided. The rate of temperature rise of the liquid when the pump is operated with a closed valve can be calculated from the equation

$$\Delta T = 42.42 \, BHP_0/(MCp) \tag{3.3}$$

where
ΔT: temperature rise, °F per min
BHP_0: BHP required under shutoff conditions
M: amount of liquid contained in pump, lb
Cp: liquid specific heat, Btu/lb/°F

In the SI units,

$$\Delta T = 59.98 \, P_0/(MCp) \tag{3.4}$$

where
ΔT: temperature rise, °C per min
P_0:   power required under shut off conditions, kW
M: amount of liquid contained in pump, kg
Cp: liquid specific heat, kJ/kg/°C

EXAMPLE 3.4 USCS UNITS

A centrifugal pump runs against a closed valve for a short period of time. The BHP curve shows that the minimum BHP at shutoff conditions is 350 HP. The pump contains 1200 lb of liquid (Cp = 0.45 Btu/lb/°F). Calculate the temperature rise per unit time.

Solution
Using Equation (3.3), we get

$$\Delta T/min = 42.42 \times 350/(1200 \times 0.45) = 27.49°F/min$$

EXAMPLE 3.5 SI UNITS

A centrifugal pump runs against a closed valve for a short period of time; the power curve shows that the minimum power at shutoff conditions is 186 kW. The pump contains 455 kg of liquid (Cp = 1.9 kJ/kg/°C). Calculate the temperature rise per unit time.

Solution
Using Equation (3.4), we get

$$\Delta T = 59.98 \times 186/(455 \times 1.9) = 12.90°C/min$$

Due to the possible excessive temperature rise at low flow rates, some minimum flow limit must be specified for a pump. Suppose a centrifugal pump has a capacity range of 0 to 2000 gal/min and the BEP is at Q = 1600, H = 1200, and E = 85%. If the pump was selected properly for the application, we would expect that the operating point would be close to and slightly to the left of the BEP. This will ensure that the pump is operating close to its best efficiency and provide for a slight increase in capacity without sacrificing efficiency. The temperature rise of the liquid when operating near the BEP may be in the range of 2° to 5°F. However, if for some reason the flow rate is cut back by using a discharge control valve, the pump pressure is throttled and the temperature rise of the liquid increases as indicated in Figure 3.6.

If the flow rate is reduced too much, excessive temperature rise would result. For volatile liquids such as gasoline or turbine fuel, this would not be acceptable. The increase in temperature may cause vaporization of the liquid, with the result that the pump will contain liquid and vapor. Since the pump is designed to handle a single-phase liquid, the existence of vapor would tend to cavitate the pump and damage the impeller. Cavitation and NPSH are discussed in detail in Chapter 7. Therefore we must set a minimum capacity limit at which a pump can be operated continuously.

The following parameters are used as a guide to determine the permissible minimum flow in a pump:

1. Temperature rise
2. Radial thrust in pump supports
3. Internal recirculation
4. Shape of the BHP curve for high-specific-speed pumps
5. Existence of entrained air or gas in the liquid

The pump vendor usually specifies the minimum flow permissible, generally ranging from between 25% and 35% of maximum pump capacity. For example, a pump with a capacity range of 0 to 3000 gpm and a BEP at Q = 2700 gpm may be restricted to a minimum flow of not less than 900 gpm.

System Head Curve

So far, we have discussed in detail pump performance curves, capacity range, range of the head, and the BEP on a pump curve. While it is desirable to operate a pump as close as possible to its BEP, it may not be always possible to do so. So how do we determine the actual operating point of a pump in an installation? The operating point on a pump curve, also called the duty point, is defined as that point on the pump H-Q curve at which the head requirement of the piping system connected to the pump exactly matches what the pump can provide at that flow rate. The piping system requires a certain amount of pressure for a certain liquid flow rate.

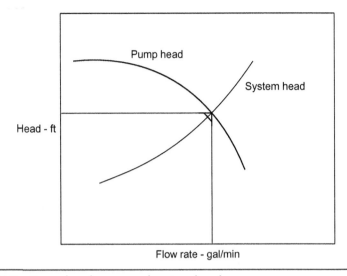

Figure 3.7 Pump head curve and system head curve.

This depends on the piping configuration, the lengths of both the suction and discharge piping, and the amount of frictional head loss in the piping system. The system head increases as the flow rate increases. The centrifugal pump head, on the other hand, decreases as the flow increases. Hence, there is a certain flow at which both the system and pump head curves match. At this flow rate the pump produces exactly what the system piping requires. This is illustrated in Figure 3.7. System head curves are discussed in more detail in the subsequent chapters. We will next review the selection of a centrifugal pump for a practical application as illustrated in Example 3.6.

Pipe Diameters and Designations

In the USCS units, steel pipe is designated by the term nominal pipe size (NPS). Thus, a 10-inch nominal pipe size is referred to as NPS 10 pipe, and it has an actual outside diameter of 10.75 in. If the wall thickness is 0.250 in., the inside diameter is 10.25 in. Similarly, NPS 12 pipe has an outside diameter of 12.75 in. For NPS 14 and above, the nominal pipe size is the same as the outside diameter. Thus, NPS 14 and NPS 20 pipes have outside diameters of 14 in. and 20 in., respectively. Refer to Appendix E for dimensional and other properties of circular pipes in USCS units.

In the SI units, the term *diametre nominal* (DN) is used instead of NPS. The actual outside diameter for DN 500 pipe is slightly larger than 500 mm. However, throughout this book, for simplicity and ease of calculation, DN 500 pipe is assumed to have an outside diameter of 500 mm. Similarly, DN 300 pipe has an outside diameter of 300 mm, and so on. Refer to Appendix F for dimensional and other properties of circular pipes in SI units.

EXAMPLE 3.6 USCS UNITS

A storage tank Tk-201 containing diesel fuel at Hartford terminal is used to pump the fuel from Hartford to another storage terminal at Compton 5 miles away, as depicted in Figure 3.8. The pump to be selected for this application will be located on a foundation approximately 20 ft away from the Hartford tank at an elevation of 120 ft above mean sea level (MSL) and connected to it via an NPS 16 pipe. On the discharge side of the pump, an NPS 12 pipe will connect the pump to a meter manifold consisting of valves and a flow meter located approximately 50 ft from the pump. From the manifold, an NPS 12 pipeline runs to the fence line of the Compton terminal and then to a meter manifold located 100 ft inside the fence. From the meter manifold, an NPS 12 pipe that is 50 ft long connects to the tank Tk-301 at the Compton terminal.

The elevation of the center line of the pump suction at Hartford is at 80 ft above MSL. Initially, the level of liquid in the tank is 30 ft above the bottom of the tank. The ground elevation of the manifold system at Hartford is 90 ft. The manifold at Compton is located at an elevation of 100 ft above MSL, and the bottom of the Compton tank has

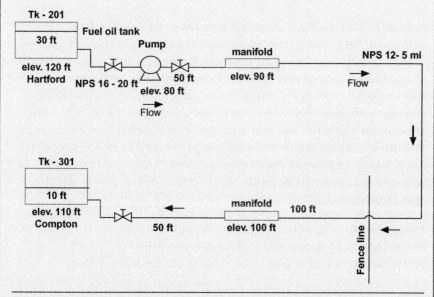

Figure 3.8 Pumping diesel fuel from Hartford to Compton.

an elevation of 110 ft above MSL. The initial liquid level at the Compton tank is 10 ft. Diesel fuel has the following properties at the operating temperature of 60°F:

Specific gravity = 0.85, and viscosity = 5.0 cSt

Assume that the meter manifolds have a fixed pressure drop of 15 psi. Select a suitable pump for this application for a diesel fuel transfer rate of 4000 bbl/h (2800 gal/min). What HP electric motor drive is required for this application?

Solution

In order to transfer diesel fuel from Tk-201 at Hartford to Tk-301 at Compton approximately 5 miles away, we need sufficient pressure at Hartford to overcome the frictional resistance in the interconnecting piping and meter manifolds. In addition, there must be sufficient pressure as the diesel arrives at the Compton tank to overcome the elevation difference between the tank bottom and the ground elevation at Compton plus the existing liquid level in Tk-301. At Hartford the free surface of the diesel fuel has an elevation of 120 + 30 ft above MSL, while the proposed pump suction has an elevation of 80 ft. Therefore, there is a positive static suction head of (150 − 80 = 70 ft) that forces the liquid into the suction of the pump. Using Equation (1.11) this static head can be converted to psi as follows:

Static suction pressure = 70 × 0.85/2.31 = 25.76 psi

In the absence of the pump, the preceding static suction pressure is obviously not enough to push the diesel fuel to Compton via the interconnecting pipeline system. In addition, as diesel flows through the 16-in. suction piping 20 ft long as shown in Figure 3.7, the available static pressure of 25.76 psi is further depleted due to the frictional resistance in the suction piping. The pump selected for this application must be capable of providing adequate pressure to move diesel at 4000 bbl/h flow rate to Compton. Therefore, we need to first determine the total frictional resistance in the piping system from the pump discharge at Hartford to the storage tank at Compton. A detailed discussion of pressure drop through pipes, valves, and fittings is covered in Chapter 4. For the present we will assume that the pressure drops for diesel flowing in 16-in. and 12-in. pipes are as follows:

Pressure drop for 16-in. pipe = 9.66 psi/mi (1.83 psi/1000 ft)
Pressure drop for 12-in. pipe = 30.43 psi/mi (5.76 psi/1000 ft)
For the total length of 12-in. pipe = 50 ft + 5 mi + 100 ft + 50 ft = 5.04 mi

Since we are not given the details of the other fittings, such as elbows, tees, other valves, and so forth, we will account for this by increasing the total length of straight pipe by

10%. (Equivalent length and pressure drops through fittings and valves are discussed in detail in Chapter 4.) Therefore, the total equivalent length of the piping system from the discharge of the pump to the tank valve at TK-301 equals $5.04 \times 1.1 = 5.54$ mi.

The total pressure drop of the discharge piping, 5.54 miles long, including the pressure drop through the two meter manifolds, is

$$5.54 \times 30.43 + 15 + 15 = 198.6 \text{ psi}$$

In addition to the preceding pressure, the pump selected must also be capable of raising the diesel from an elevation of 80 ft at the pump discharge to the tank level of $110 + 10 = 120$ ft:

$$\text{Static discharge head} = 110 + 10 - 80 = 40 \text{ ft}$$

Therefore, the minimum discharge pressure required of the pump is

$$198.6 + (40 \times 0.85 / 2.31) = 213.32 \text{ psi approximately}$$

On the suction side of the pump we estimated the static suction pressure to be 25.76 psi. The actual suction pressure at the pump is obtained by reducing static suction pressure by the pressure drop in the 16-in. suction piping as follows:

$$\text{Pump suction pressure} = 25.76 - ((9.66 \times 20 \times 1.5)/5280) = 25.70 \text{ psi}$$

Notice that we increase the length of the suction piping by 50% to account for valves and fittings. Therefore, the pump we need must be able to produce a differential pressure of

$$\Delta P = (213.32 - 25.7) = 187.62 \text{ psi}$$

Converting pressure to ft of head of diesel, the pump differential head required is

$$\Delta H = 187.62 \times 2.31 / 0.85 = 510 \text{ ft}$$

Therefore, the pump selected should have the following specifications:

$$Q = 2800 \text{gal/min}, H = 510 \text{ ft with an efficiency of 80\% to 85\%}$$

From a manufacturer's catalog we will choose a pump that meets the preceding conditions at its BEP. In order to ensure that there is adequate suction head, we must also calculate the available NPSH and compare that with the NPSH required from the manufacturer's pump performance data. NPSH will be discussed in Chapter 7.

In order to determine the drive motor HP for this pump, we must know the efficiency of the pump under these conditions. Assuming 80% pump efficiency, using Equation (2.5), the BHP required under these conditions is

$$BHP = 2800 \times 510 \times 0.85/(3960 \times 0.8) = 383 \text{ HP } (286 \text{ kW})$$

This is the minimum HP required at the operating condition of 4000 bbl/h (2800 gal/min). The actual installed HP of the motor must be sufficient to handle the BHP requirement at the maximum capacity of the pump curve. Increasing the preceding by 10% and considering 95% motor efficiency,

The drive motor HP required is

$$\text{Motor HP} = 383 \times 1.1/0.95 = 444 \text{ HP}$$

Choosing the nearest standard size motor, a 500-HP motor is recommended for this application.

EXAMPLE 3.7 SI UNITS

Fuel oil contained in a storage tank Tk-105 at Salinas terminal needs to be pumped from Salinas to another storage terminal at Fontana 12 km away. The pump at Salinas will be located on a foundation at an elevation of 50 m above mean sea level (MSL) and 10 m away from the storage tank, connected to the tank by a pipe DN 500, 10 mm wall thickness. On the discharge side of the pump, a pipe DN 300, 8 mm wall thickness will connect the pump to a meter manifold consisting of valves and a flow meter located approximately 20 m from the pump, as shown in Figure 3.9.

From the manifold a DN 300 pipeline, 12 km long runs to the fence line of Fontana terminal and from there to a meter manifold located 40 m inside the fence. From the meter manifold a DN 300 pipe, 20 m long connects to the tank Tk-504 at the Fontana terminal. The elevation of the centerline of the suction of the pump at Salinas is at 30 m above MSL. Initially, the level of liquid in the Salinas tank is 10 m above the bottom of the tank. The ground elevation of the manifold system at Salinas is 29 m. The manifold at Fontana is located at an elevation of 32 m above MSL, and the bottom of the Fontana tank has an elevation of 40 m above MSL.

The initial liquid level at both tanks is 10 m. The fuel oil has the following properties at the operating temperature of 20°C:

$$\text{Specific gravity} = 0.865 \text{ and viscosity} = 5.8 \text{ cSt}$$

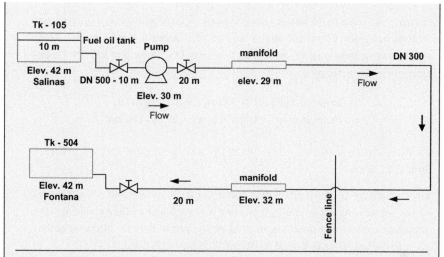

Figure 3.9 Pumping fuel oil from Salinas to Fontana.

Assume that the meter manifolds have a fixed pressure drop of 1.0 bar. Select a suitable pump for this application for a fuel oil transfer rate of 210 L/s. What is the power required for the electric motor drive for this pump?

Solution

In order to transfer fuel oil from Tk-105 at Salinas to Tk-504 at Fontana 12 km away, we need sufficient pressure at Salinas to overcome the frictional resistance in the interconnecting piping and meter manifolds. In addition, there must be sufficient pressure as the product arrives at the Fontana tank to overcome the elevation difference between the Salinas tank and the ground elevation at Fontana plus the existing liquid level in Tk-504. At Salinas the free surface of the fuel oil is at an elevation of $(42 + 10)$ m above MSL, while the proposed pump suction has an elevation of 30 m. Therefore, there is a positive static suction head of $52 - 30 = 22$ m that forces the liquid into the suction of the pump. Using Equation (1.12) this static head can be converted to kPa as follows:

$$\text{Static suction pressure} = 22 \times 0.865 / 0.102 = 186.57 \text{ kPa}$$

In the absence of the pump, the preceding static suction pressure is obviously not enough to push the fuel oil to Fontana via the interconnecting pipeline system. In addition, as fuel oil flows through the DN 500 suction piping 10 m long as shown in Figure 3.9, the available static pressure of 186.57 kPa is further depleted due to the frictional resistance in the suction piping. The pump selected for this application must be capable of providing adequate pressure to move fuel oil at 210 L/s flow rate to Fontana. Therefore, we need to first determine the total frictional resistance in the piping system from the pump discharge at Salinas to the storage tank at Fontana. A detailed discussion of pressure drop through pipes, valves, and fittings is covered in Chapter 4. For the present, we will assume that the pressure drops for fuel oil flowing in DN 500 suction and DN 300 discharge pipes are as follows:

$$\text{Pressure drop for DN 500 pipe} = 23.15 \text{ kPa/km}$$
$$\text{Pressure drop for DN 300 pipe} = 297.34 \text{ kPa/km}$$

For the discharge piping, the total length of DN 300 pipe = 20 m + 12 km + 40 m + 20 m = 12.08 km

Since we are not given the details of the other fittings, such as elbows, tees, and other valves, we will account for this by increasing the total length of straight pipe by 10%. Therefore, the total equivalent length of the piping system from the discharge of the pump to the tank valve at Tk-504 equals 12.08 × 1.1 = 13.29 km.

The total pressure drop of the discharge piping 13.29 km long, including the pressure drop through the two meter manifolds, is

$$(13.29 \times 297.34) \text{ kPa} + (1.0 + 1.0) \text{ bar} = 4151.65 \text{ kPa}$$

In addition to the preceding pressure, the pump selected must also be capable of raising the fuel oil from an elevation of 30 m at the pump discharge to the Tk-504 level of 40 + 10 = 50 m:

$$\text{Static discharge head} = 50 - 30 = 20 \text{ m}$$

Therefore, the minimum discharge pressure required of the pump is

$$4151.65 \text{ kPa} + (20 \times 0.865/0.102) \text{ kPa} = 4321.26 \text{ kPa}$$

On the suction side of the pump, we estimated the static suction pressure to be 186.57 kPa. The actual suction pressure at the pump is obtained by reducing static suction pressure by the pressure drop in the DN 500 suction piping as follows:

$$\text{Pump suction pressure} = 186.57 - (23.15 \times 10 \times 1.5)/1000 = 186.22 \text{ kPa}$$

Notice that we increased the length of the suction piping by 50% to account for valves and fittings. Therefore, the pump we need must be able to produce a differential pressure of

$$\Delta P = (4321.26 - 186.22) = 4135.04 \text{ kPa}$$

Converting this pressure to m of head of fuel oil, the pump differential head required is

$$\Delta H = 4135.04 \times 0.102/0.865 = 488 \text{ m}$$

Therefore, the pump selected should have the following specifications:

$$Q = 210 \text{ L/s}, H = 488 \text{ m}$$

From a manufacturer's catalog we will choose a pump that meets the preceding conditions at its BEP. In order to ensure that there is adequate suction head, we must also calculate the available NPSH and compare that with the NPSH required from the manufacturer's pump performance data. NPSH calculations are discussed in Chapter 7.

In order to determine the motor power for this pump, we must know the efficiency of the pump under these conditions. Assuming 80% pump efficiency, using Equation (2.6) the power required under these conditions is

$$\text{Power} = 210 \times 3600/1000 \times 488 \times 0.865/(367.46 \times 0.8) = 1086 \text{ kW}$$

This is the minimum pump power required at the operating condition of 210 L/s. The actual installed power of the motor must be sufficient to compensate for the motor efficiency (around 95%) and handle the power requirement at the maximum capacity of the pump curve. Increasing the preceding by 10% and choosing the nearest standard size motor, a 1300 kW motor is recommended for this application.

Summary

In this chapter we discussed the performance of a pump and how it is affected by the specific gravity and viscosity of the liquid pumped. The degradation of the head and efficiency when pumping high viscosity liquids was reviewed. The Hydraulic Institute chart method of determining the viscosity corrected pump performance was explained in detail. The use of popular software PUMPCALC (*www.systek.us*) for determining the performance of a pump with high viscosity liquids was reviewed. The temperature rise of a liquid due to the pump inefficiency was explained and the danger of running a pump with a closed discharge valve was illustrated using an example. The concept of minimum pump flow was discussed. Finally, an example of how a pump is selected for a particular application was explained. This example illustrated how the pump head was calculated for transporting diesel fuel from one storage location to another storage terminal.

Problems

3.1 A 5-stage centrifugal pump has the following H-Q and E-Q data taken from the pump curve with water as the liquid pumped. Determine the viscosity corrected performance when pumping crude oil with Sg = 0.895 and viscosity = 300 cSt at 60°F.

Q gpm	600	1200	2400	4000	4500
H ft	2520	2480	2100	1680	1440
E %	34.5	55.7	79.3	76.0	72.0

What is the maximum BHP and select a suitable electric motor drive?

3.2 For a centrifugal pump the viscosity correction factors at BEP are as follows:

$$C_E = 0.59 \qquad C_Q = 0.94 \qquad C_H = 0.90$$

The BEP for water performance is Q = 2400 L/min, H = 620 m, and E = 82%.
   Compare the power required when pumping water and the viscous liquid with Sg = 0.95 and viscosity = 250 cSt at 15°C.

3.3 Calculate the temperature rise of liquid due to pumping at the following conditions

$$Q = 1200 \, gpm, \; H = 1800 \, ft, \; E = 79.5\%$$

The liquid has a specific heat of 0.44 Btu/lb/°F.

3.4 A centrifugal pump is operated for a short period of time against a closed discharge valve. Its power curve at shutoff indicates 120 kW. The pump contains 504 kg of liquid with a specific heat of Cp = 1.9 kJ/kg/°C. Calculate the temperature rise per unit time.

3.5 The static suction head on a pump is 12 m, while the discharge head is 22 m. The suction piping is DN 400, 8 mm wall thickness, and the discharge piping is DN 300, 8 mm wall thickness. The total equivalent lengths of the suction and discharge piping are 8.4 m and 363.5 m, respectively. Develop an equation for the system head curve as a function of capacity Q in m³/h.

Chapter 4

Pressure Loss through Piping Systems

In Chapter 3, to select a pump, we assumed a certain pressure drop through the piping system in order to calculate the system head requirements. In this chapter we will review the basic concepts of friction loss in pipes and explain how to calculate the head loss using the Darcy equation and the Colebrook friction factor. The popular head loss formula Hazen-Williams will also be introduced in relation to water pumping systems. The Moody Diagram method as an alternative to the Colebrook equation for friction factor determination will be illustrated using examples. We will also address the minor losses associated with fittings and valves and the concept of the equivalent length method. Series and parallel piping systems and calculation of equivalent diameters will be explained with examples.

DOI: 10.1016/B978-1-85617-828-0.00004-4

Velocity of Flow

Consider the flow of a liquid in a circular pipe with an inside diameter D. The average velocity of flow can be calculated using the formula

$$V = \text{volume flow rate/area of flow} \tag{4.1}$$

If the volume flow rate is Q ft^3/s, the velocity is

$$V = Q \times 144/(\pi\, D^2/4) \tag{4.2}$$

where
 V: average flow velocity, ft/s
 Q: flow rate, ft^3/s
 D: inside diameter of the pipe, inches
Simplifying, we can write the equation as

$$V = (\text{Const})\, Q/D^2 \tag{4.3}$$

where C is a constant depending on the units chosen.

 This basic equation for velocity of flow can be modified for use with the more common units used in the industry as follows:
 In USCS units:

$$V = 0.4085\, Q/D^2 \tag{4.4}$$

where
 V: average flow velocity, ft/s
 Q: flow rate, gal/min
 D: inside diameter of pipe, inches

$$V = 0.2859\, Q/D^2 \tag{4.5}$$

where
 V: average flow velocity, ft/s
 Q: flow rate, bbl/h
 D: inside diameter of pipe, inches
 In SI units, velocity is calculated as follows:

$$V = 353.6777\, Q/D^2 \tag{4.6}$$

where
 V: average flow velocity, m/s

Q: flow rate, m^3/h

D: inside diameter of pipe, mm

$$V = 1273.242 \, Q / D^2 \qquad (4.6a)$$

where

V: average flow velocity, m/s

Q: flow rate, L/s

D: inside diameter of pipe, mm

EXAMPLE 4.1 USCS UNITS

Water flows through a 16-inch pipe with a wall thickness of 0.250 inch at the rate of 3000 gal/min. What is the average flow velocity?

Solution

The inside diameter of pipe D = 16 − (2 × 0.250) = 15.5 in. Using Equation (4.4), the average velocity in ft/s is

$$V = 0.4085 \times 3000/(15.5)^2 = 5.1 \text{ft/s}$$

EXAMPLE 4.2 SI UNITS

Gasoline flows through the discharge piping system at a flow rate of 800 m^3/h. The pipe is 500 mm outside diameter and 10 mm wall thickness. Calculate the velocity of flow.

Solution

The inside diameter of pipe D = 500 − (2 × 10) = 480 mm. Using Equation (4.6), the average velocity in m/s is

$$V = 353.6777 \times 800/(480)^2 = 1.23 \text{ m/s}$$

In the preceding calculation, the average velocity was calculated based on the flow rate and inside diameter of pipe. As explained in Chapter 1, Figure 1.4, the shape of the velocity profile at any cross section of a pipe depends on the type of flow. It may be parabolic or trapezoidal in shape. At low flow rates and under *laminar flow* conditions, the velocity profile approximates a parabola. At higher flow rates and in *turbulent flow*, the velocity profile approximates a trapezoidal shape.

Types of Flow

Flow through a pipe may be classified as laminar, turbulent, or critical flow. The parameter, called the *Reynolds number*, is used to determine the type of flow. The

Reynolds number of flow is a dimensionless parameter that is a function of the pipe diameter, flow velocity, and liquid viscosity. It can be calculated as follows:

$$R = VD/\nu \qquad (4.7)$$

where

R: Reynolds number, dimensionless parameter
V: average flow velocity
D: inside diameter of pipe
ν: liquid viscosity

The units for V, D, and ν are chosen such that R is a dimensionless term. For example, V in ft/s, D in ft, and ν in ft²/s will result in R being dimensionless, or without units. Similarly, in SI units, V in m/s, D in m, and ν in m²/s will make R dimensionless. More convenient forms for R using common units will be introduced later in this chapter.

Once the Reynolds number of flow is known, the flow can be classified as follows:

Laminar flow	$R \leq 2000$
Critical flow	$R > 2000$ and $R < 4000$
Turbulent flow	$R \geq 4000$

In some publications, the upper limit for laminar flow may be stated as 2100 instead of 2000. Laminar flow derives its name from *lamination*, indicating that such a flow results in laminations of smoothly flowing streams of liquid in the pipes. Laminar flow is also known as viscous flow, in which eddies or turbulence do not exist.

When the flow rate is increased, turbulence and eddies are created due to friction between the liquid and pipe wall. For Reynolds number values between 2000 and 4000, critical flow is said to exist. When the Reynolds number is greater than 4000, more turbulence and eddies are created, and flow is classified as fully turbulent flow. The critical flow zone that exists between laminar and turbulent flow is also referred to as an unstable region of flow. These flow regions are graphically represented in Figure 4.1, which shows the range of Reynolds numbers for each flow regime.

Laminar (or viscous) flow is found to occur with high-viscosity fluids and at low flow rates. Turbulent flow generally occurs with higher flow rates and lower viscosity. Using customary units for pipe diameter, viscosity, and flow rate instead of velocity, the Reynolds number in Equation (4.7) becomes the following:
In USCS units:

$$R = 3160\, Q/(\nu D) \qquad (4.8)$$

where

R: Reynolds number, dimensionless
Q: flow rate, gal/min

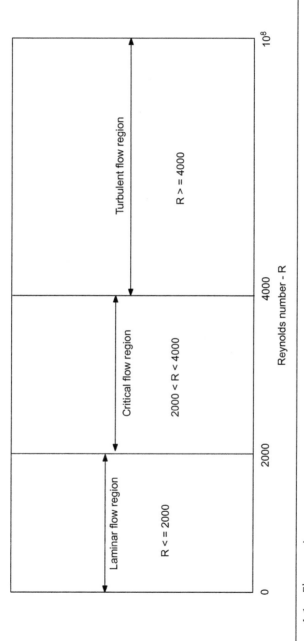

Figure 4.1   Flow regions.

v: kinematic viscosity of the liquid, cSt

D: inside diameter of pipe, inch

$$R = 2214 \, Q/(vD) \tag{4.9}$$

where

R: Reynolds number, dimensionless

Q: flow rate, bbl/h

v: kinematic viscosity of the liquid, cSt

D: inside diameter of pipe, inch

In SI units:

The Reynolds number is calculated as follows:

$$R = 353,678 \, Q/(vD) \tag{4.10}$$

where

R: Reynolds number, dimensionless

Q: flow rate, m^3/h

v: kinematic viscosity of the liquid, cSt

D: inside diameter of pipe, mm

$$R = 1.2732 \times 10^6 \, Q/(vD) \tag{4.10a}$$

where

R: Reynolds number, dimensionless

Q: flow rate, L/s

v: kinematic viscosity of the liquid, cSt

D: inside diameter of pipe, mm

EXAMPLE 4.3 USCS UNITS

Calculate the Reynolds number of flow for a crude oil pipeline that is 16 in. outside diameter and 0.250 in. wall thickness at a flow rate of 6250 bbl/h. Viscosity of the crude oil is 15.0 cSt.

Solution

The inside diameter of pipe D = 16 − (2 × 0.250) = 15.5 inches. Using Equation (4.9), we calculate the Reynolds number as follows:

$$R = (2214 \times 6250)/(15.5 \times 15) = 59,516$$

EXAMPLE 4.4 SI UNITS

Diesel (viscosity = 5.0 cSt) flows through a 400 mm outside diameter pipeline, 8 mm wall thickness at a flow rate of 150 L/s. Determine the Reynolds number of flow. At what range of flow rates will the flow be in the critical zone?

Solution

Flow rate Q = 150 L/s

Pipe inside diameter D = 400 − (2 × 8) = 384 mm

Using Equation (4.10a), we calculate the Reynolds number as follows:

$$R = (1.2732 \times 10^6 \times 150)/(5 \times 384) = 99{,}469$$

Since R is greater than 4000, the flow is turbulent.

For the lower limit of the critical zone:

$$(1.2732 \times 10^6 \times Q)/(5 \times 384) = 2000$$

Solving for Q, we get Q = 3.02 L/s. The upper limit occurs at when R = 4000, and the corresponding flow rate is 6.03 L/s.

EXAMPLE 4.5 SI UNITS

Heavy crude oil with a viscosity of 650 cSt flows through a 500 mm diameter pipe, 10 mm wall thickness. What minimum flow rate is required to ensure turbulent flow?

Solution

Pipe inside diameter D = 500 − (2 × 10) = 480 mm

Using Equation (4.10), the Reynolds number at a flow rate of Q m^3/h is

$$(353{,}678 \times Q)/(650 \times 480) = 4000 \text{ for turbulent flow}$$

Solving for Q, we get

$$Q = 3529 \text{ m}^3/\text{h}$$

Pressure Drop Due to Friction

As liquid flows through a pipe, due to viscosity, the frictional resistance between the liquid and the pipe wall causes energy loss or pressure drop (or head loss) from the upstream end to the downstream end of the pipe. The amount of pressure drop per unit length of pipe is a function of the flow velocity, inside diameter of the pipe, and the specific gravity of the liquid. The head loss due to friction in a given length and diameter of pipe at a certain liquid flow velocity can be calculated using the Darcy equation:

$$h = f(L/D)V^2/2g \qquad (4.11)$$

where
    h: head loss due to friction, ft
    f: friction factor, dimensionless
    L: pipe length, ft
    D: inside diameter of pipe, ft
    V: average flow velocity, ft/s
    g: acceleration due to gravity $= 32.2\,\text{ft/s}^2$
    In SI units, the Darcy equation is as follows:

$$h = f(L/D)V^2/2g \qquad (4.11a)$$

where
    h: head loss due to friction, m
    f: friction factor, dimensionless
    L: pipe length, m
    D: inside diameter of pipe, m
    V: average flow velocity, m/s
    g: acceleration due to gravity $= 9.81\,\text{m/s}^2$
The Darcy equation is also referred to as the Darcy-Weisbach equation for pressure drop due to friction.

The friction factor f (also known as the Darcy friction factor) depends on the internal roughness of the pipe, the inside diameter of pipe, and the Reynolds number for turbulent flow. For laminar flow, f depends only on the Reynolds number. The values range from 0.008 to 0.10, as shown in the Moody diagram in Figure 4.2. Sometimes the Darcy friction factor is also referred to as the Moody friction factor, since it can be read off the Moody diagram. In some publications you may see the term *Fanning friction factor*. This is simply one-fourth the value of the Darcy (or Moody) friction factor, as follows:

$$\text{Fanning friction factor} = \text{Darcy friction factor}/4$$

In this book, we will use the Darcy (or Moody) friction factor only.

The relationship between the friction factor f and Reynolds number R for various flow regimes (laminar, critical, and turbulent) are as shown in Figure 4.2. In

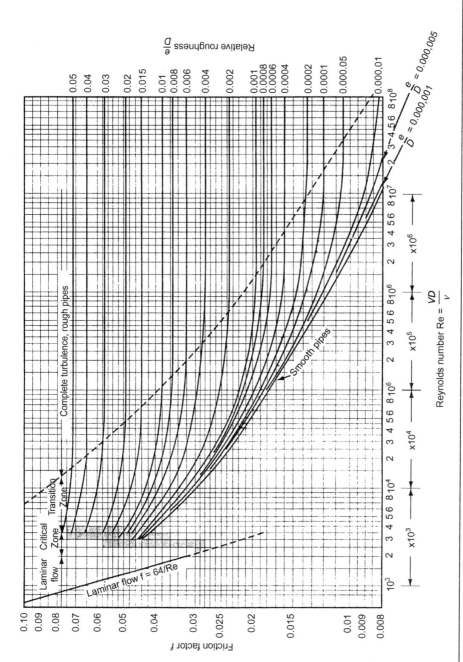

Figure 4.2 Moody diagram for friction factor.

turbulent flow, the friction factor f is also a function of the relative pipe roughness, a dimensionless parameter obtained by dividing the absolute internal pipe roughness (e) by the inside diameter (D) of the pipe.

Examination of the Moody diagram shows that in laminar flow, for Reynolds numbers less than 2000, the friction factor decreases as the flow rate increases, reaching a value of approximately f = 0.032 at the boundary value of R = 2000.

In the turbulent zone, for R > 4000, the friction factor depends on both the value of R as well as the relative roughness of pipe (e/D). As R increases in value beyond 10 million or so, f is a function of the relative roughness alone. This range called the complete turbulence in rough pipes and is designated as the Moody diagram region to the right of the dashed line. The portion between the critical zone and the dashed line is called the transition zone. In this zone the influence of Reynolds number on friction factor is more pronounced. For example, the Moody diagram shows that for R = 100,000 and relative roughness of 0.001, the friction factor is approximately 0.0225.

In the critical zone (between R = 2000 and R = 4000), the value of the friction factor is undefined. Some empirical correlations have been put forth to cover this range of Reynolds numbers, but most of the time, the turbulent zone friction factor is used to be conservative. The relative roughness of the pipe is defined as

Relative roughness = Absolute roughness/inside diameter of pipe = e/D

For new steel pipe, the absolute roughness of e = 0.0018 inch may be used. Table 4.1 lists other values used for absolute roughness for typical pipe materials.

Table 4.1 Pipe roughness

Pipe Material	Pipe Roughness in	Pipe Roughness mm
Riveted steel	0.0354 to 0.354	0.9 to 9.0
Concrete	0.0118 to 0.118	0.3 to 3.0
Wood stave	0.0071 to 0.0354	0.18 to 0.9
Cast iron	0.0102	0.26
Galvanized iron	0.0059	0.15
Asphalted cast iron	0.0047	0.12
Commercial steel	0.0018	0.045
Wrought iron	0.0018	0.045
Drawn tubing	0.000059	0.0015

For example, if the absolute roughness of pipe is 0.0018 inch, and the inside diameter of pipe is 15.5 inch, the relative roughness is

$$e/D = 0.0018/15.5 = 0.00012$$

Being a ratio of similar units, the relative roughness is a dimensionless term. In SI units, e and D are stated in mm, so the ratio is dimensionless.

EXAMPLE 4.6 USCS UNITS

Water flows through a new steel pipe 14 inches diameter and 0.250 inch wall thickness at a flow rate of 3000 gal/min. If the Darcy friction factor f = 0.025, calculate the head loss due to friction for 1000 ft of pipe.

Solution

Inside diameter of pipe D = 14 − (2 × 0.250) = 13.5 inch. Using Equation (4.4), the velocity is

$$V = 0.4085 \times 3000/(13.5)^2 = 6.72 \text{ ft/s}$$

From Equation (4.11), the head loss due to friction is

$$h = 0.025(1000 \times 12/13.5)(6.72)^2/64.4 = 15.58 \text{ ft}$$

Converting to pressure loss in psi, the pressure drop = 15.58 × 1.0/2.31 = 6.75 psi.

EXAMPLE 4.7 SI UNITS

Water flows in a 450 mm diameter pipe, 8 mm wall thickness, at a flow rate of 200 L/s. Calculate the head loss due to friction for a pipe with a length of 500 m. Assume f = 0.02.

Solution

The inside diameter of pipe D = 450 − (2 × 8) = 434 mm. Using Equation (4.6), the velocity is

$$V = 353.6777 \times (0.2 \times 3600)/(434)^2 = 1.352 \text{ m/s}$$

From Equation (4.11a), the head loss due to friction is

$$h = 0.02 \, (500/0.434)/(1.352)^2/(2 \times 9.81) = 2.147 \text{ m}$$

Converting to pressure loss in kPa, the pressure drop = 2.147 × 1.0/0.102 = 21.05 kPa.

In the previous examples, we assumed a value of the friction factor. Shortly, we will explain how f can be determined from the Moody diagram when you know the Reynolds number and the pipe roughness. The Darcy equation as stated in Equation (4.11) is the classic equation for the frictional pressure drop calculation. Using conventional units and the flow rate instead of the velocity, we have a modified pressure drop equation that is easier to use, as described next.

In USCS units:

The pressure drop P_m in psi/mi is calculated from the flow rate Q, liquid specific gravity, and pipe inside diameter D as follows:

$$P_m = 71.1475 \ fQ^2 Sg/D^5 \tag{4.12}$$

$$h_m = 164.351 \ fQ^2/D^5 \tag{4.12a}$$

where

P_m: pressure drop, psi/mi
h_m: head loss, ft of liquid/mi
f: friction factor, dimensionless
Q: flow rate, gal/min
Sg: specific gravity of liquid, dimensionless
D: inside diameter of pipe, inch

When flow rate is in bbl/h, the equation becomes

$$P_m = 34.8625 \ fQ^2 Sg/D^5 \tag{4.13}$$

$$h_m = 80.532 \ fQ^2/D^5 \tag{4.13a}$$

where

P_m: pressure drop, psi/mi
h_m: head loss, ft of liquid/mi
f: friction factor, dimensionless
Q: flow rate, bbl/h
Sg: specific gravity of liquid, dimensionless
D: inside diameter of pipe, inch

In SI units:

In SI units, the pressure drop due to friction is stated in kPa/km and is calculated as follows:

$$P_{km} = 6.2475 \times 10^{10} \ f Q^2 \ (Sg/D^5) \tag{4.14}$$

$$h_{km} = 6.372 \times 10^9 \ f Q^2 \ (1/D^5) \tag{4.14a}$$

where

P_{km}: pressure drop, kPa/km
h_{km}: head loss, m of liquid/km
f: friction factor, dimensionless
Q: flow rate, m^3/h
Sg: specific gravity of liquid, dimensionless
D: inside diameter of pipe, mm

Sometimes the term F, which is called the transmission factor, is used instead of the friction factor f in the pressure drop equations. These two terms are inversely related:

$$F = 2/\sqrt{f} \qquad (4.15)$$

Alternatively,

$$f = 4/F^2 \qquad (4.16)$$

In terms of the transmission factor F, the pressure drop in Equations (4.15) and (4.16) may be restated as follows:
In USCS units:

$$P_m = 284.59 \, (Q/F)^2 \, Sg/D^5 \qquad \text{for Q in gal/min} \qquad (4.17)$$

$$P_m = 139.45 \, (Q/F)^2 \, Sg/D^5 \qquad \text{for Q in bbl/h} \qquad (4.18)$$

In SI units:

$$P_{km} = 24.99 \times 10^{10} \, (Q/F)^2 \, (Sg/D^5) \qquad \text{for Q in } m^3/hr \qquad (4.19)$$

Note that since the transmission factor F and the friction factor f are inversely related, a higher transmission factor means a lower friction factor and a higher flow rate Q. Conversely, the higher the friction factor, the lower the transmission factor and flow rate. Thus, the flow rate is directly proportional to F and inversely proportional to f.

Determining the Friction Factor from the Moody Diagram

The friction factor f can be determined using the Moody diagram shown in Figure 4.2 as follows:

1. For the given flow rate, liquid properties, and pipe size, calculate the Reynolds number of flow using Equation (4.8).

2. Calculate the relative roughness (e/D) of the pipe by dividing the pipe absolute roughness by the inside diameter of the pipe.

3. Starting at the Reynolds number value on the horizontal axis of the Moody diagram, Figure 4.2, move vertically up to the relative roughness curve. Then move horizontally to the left and read the friction factor f on the vertical axis on the left.

EXAMPLE 4.8 USCS UNITS

Using the Moody diagram, determine the friction factor for a crude oil pipeline with a 16-inch outside diameter and a 0.250-inch wall thickness at a flow rate of 6250 bbl/h. Viscosity of the crude oil is 15.0 cSt. The absolute pipe roughness = 0.002 in.

Solution

The inside diameter of pipe D = 16 − (2 × 0.250) = 15.5 in. Using Equation 4.9, we calculate the Reynolds number as follows:

$$R = (2214 \times 6250)/(15.5 \times 15) = 59{,}516$$

Relative roughness = 0.002/15.5 = 0.000129. From the Moody diagram, for R = 59,516 and (e/D) = 0.000129, we get the friction factor as f = 0.0206.

EXAMPLE 4.9 SI UNITS

Using the Moody diagram, determine the friction factor for a water pipeline with a 400 mm outside diameter and a 6 mm wall thickness at a flow rate of 400 m³/h. Viscosity of water is 1.0 cSt. The absolute pipe roughness = 0.05 mm.

Solution

The inside diameter of pipe D = 400 − (2 × 6) = 388 mm. Using Equation 4.10, we calculate the Reynolds number as follows:

$$R = (353{,}678 \times 400)/(1 \times 388) = 364{,}617$$

Relative roughness = 0.05/388 = 0.000129. From the Moody diagram, for R = 364,617 and (e/D) = 0.000129, we get the friction factor as f = 0.0153.

Calculating the Friction Factor: the Colebrook Equation

We have seen how the friction factor f is determined using the Moody diagram. Instead of this, we can calculate the friction factor using the Colebrook (sometimes known as Colebrook-White) equation for turbulent flow. For laminar flow, the friction factor is calculated using the simple equation

$$f = 64/R \tag{4.20}$$

For turbulent flow, the Colebrook-White equation for friction factor is

$$1/\sqrt{f} = -2 \, \text{Log}_{10}[(e/3.7D) + 2.51/(R\sqrt{f})] \tag{4.21}$$

where

f: friction factor, dimensionless

D: inside diameter of pipe, in.

e: absolute roughness of pipe, in.

R: Reynolds number, dimensionless

In SI units, Equation (4.21) can be used if D and e are both in mm. R and f are dimensionless.

Since the friction factor f appears on both sides of Equation (4.21), the calculation of f must be done using a trial-and-error approach. Initially, we assume a value for f (such as 0.02) and substitute the values into the right-hand side of Equation (4.21). A second approximation for f is then calculated. This value can then be used on the right-hand side of the equation to obtain the next better approximation for f, and so on. The iteration is terminated, when successive values of f are within a small value such as 0.001. Usually three or four iterations are sufficient.

As mentioned before, for the critical flow region ($2000 < R < 4000$), the friction factor is considered undefined, and the turbulent friction factor is used instead. There have been several correlations proposed for the critical zone friction factor in recent years. Generally, it is sufficient to use the turbulent flow friction factor in most cases when the flow is in the critical zone.

The calculation of the friction factor f for turbulent flow using the Colebrook-White equation requires an iterative approach, since the equation is an implicit one. Due to this, many explicit equations have been proposed by researchers that are much easier to use than the Colebrook-White equation, and these have been found to be quite accurate compared to the results of the Moody diagram. The Swamee-Jain equation or the Churchill equation for friction factor may be used instead of the Colebrook-White equation, as described next.

Explicit Equations for the Friction Factor

P. K. Swamee and A. K. Jain proposed an explicit equation for the friction factor in 1976 in the *Journal of the Hydraulics Division of ASCE*. The Swamee-Jain equation is as follows:

$$f = 0.25 / [\text{Log}_{10}(e/3.7D + 5.74/R^{0.9})]^2 \tag{4.22}$$

Another explicit equation for the friction factor, proposed by Stuart Churchill, was reported in *Chemical Engineering* in November 1977. It requires the calculation of parameters A and B, which are functions of the Reynolds number R. Churchill's equation for f is as follows:

$$f = [(8/R)^{12} + 1/(A + B)^{3/2}]^{1/12} \qquad (4.23)$$

where parameters A and B are defined as

$$A = [2.457 \text{Log}_e(1/((7/R)^{0.9} + (0.27e/D))]^{16} \qquad (4.24)$$

$$B = (37\,530/R)^{16} \qquad (4.25)$$

EXAMPLE 4.10 USCS UNITS

Using the Colebrook-White equation, calculate the friction factor for flow in an NPS 20 pipe with a 0.500-in. wall thickness. The Reynolds number is 50,000, and assume an internal pipe roughness of 0.002 in. Compare the value of f obtained with the Swamee-Jain equation.

Solution

The inside diameter is

$$D = 20 - 2 \times 0.500 = 19.0 \text{ inch}$$

Using Equation (4.21) for turbulent flow friction factor:

$$1/\sqrt{f} = -2\,\text{Log}_{10}[(0.002/(3.7 \times 19.0)) + 2.51/(50000)\sqrt{f}]$$

This implicit equation in f must be solved by trial and error.

Initially, assume f = 0.02, and calculate the next approximation as

$$1/\sqrt{f} = -2\,\text{Log}_{10}[(0.002/(3.7 \times 19.0)) + 2.51/(50000\sqrt{0.02})] = 6.8327$$

Solving, f = 0.0214.

Using this as the second approximation, we calculate the next approximation for f as

$$1/\sqrt{f} = -2\,\text{Log}_{10}[(0.002/3.7 \times 19.0) + 2.51/(50000\sqrt{0.0214})] = 6.8602$$

Solving, f = 0.0212, which is close enough.

Next, we calculate f using the Swamee-Jain equation (4.22):

$$f = 0.25/[\text{Log}_{10}(0.002/(3.7 \times 19.0)) + (5.74/50,000^{0.9})]^2 = 0.0212,$$

which is the same as what we got using the Colebrook-White equation.

It can be seen from Equations (4.22) and (4.23) that the calculation of the friction factor does not require a trial-and-error approach, unlike the Colebrook-White equation. Appendix H lists the Darcy friction factors for a range of Reynolds numbers and relative roughness values. The friction factors were calculated using the Swamee-Jain equation.

Hazen-Williams Equation for Pressure Drop

Although the Moody diagram and the Colebrook-White equation are in popular use for pressure drop calculations, the water industry has traditionally used the Hazen-Williams equation. Recently, the Hazen-Williams equation has also found use in pressure drop calculations in refined petroleum products, such as gasoline and diesel. The Hazen-Williams equation can be used to calculate the pressure drop in a water pipeline from a given pipe diameter, a flow rate, and a C factor that takes into account the internal condition of the pipe. The dimensionless parameter C is called the Hazen-Williams C factor and depends on the internal roughness of the pipe. Unlike the friction factor in the Colebrook-White equation, the C factor increases with the smoothness of the pipe and decreases with an increase in pipe roughness. Therefore, the C factor is more like the transmission factor F discussed earlier. Typical values of C factors range from 60 to 150, depending on the pipe material and roughness as listed in Table 4.2.

When used with water pipelines, a C value of 100 or 120 may be used. With gasoline, a C value of 150 is used, whereas 125 is used for diesel. Note that these are simply approximate values to use when better data are not available. Generally, the value of C used is based on experience with the particular liquid and pipeline. Hence, C varies from pipeline to pipeline and with the liquid pumped. In

Table 4.2 Hazen-Williams C factors

Pipe Material	C-Factor
Smooth Pipes (All metals)	130–140
Smooth Wood	120
Smooth Masonry	120
Vitrified Clay	110
Cast Iron (Old)	100
Iron (worn/pitted)	60–80
Polyvinyl Chloride (PVC)	150
Brick	100

comparison, the Colebrook-White equation is used universally for all types of liquids. A comparison with the Hazen-Williams equation can be made to ensure that the value of C used in the Hazen-Williams equation is not too far off. Historically, the Hazen-Williams equation has been found to be accurate for water pipelines at room temperatures, for conventional velocities, and in transition flow. Considerable discrepancies were found when they were used with extreme velocities and hot- and cold-water pipelines. Sometimes a range of C values is used to estimate the flow rate and the head loss.

The most common form of the Hazen-Williams equation for pressure drop in water pipelines is as follows:

$$h = 4.73\, L(Q/C)^{1.852}/D^{4.87} \tag{4.26}$$

where
   h: head loss due to friction, ft
   L: pipe length, ft
   Q: flow rate, ft^3/s
   C: Hazen-Williams C factor
   D: inside diameter of pipe, ft
In commonly used units, another form of Hazen-Williams equation is as follows:

$$Q = 6.7547 \times 10^{-3}\, (C)\, (D)^{2.63}\, (h)^{0.54} \tag{4.27}$$

where
   Q: flow rate, gal/min
   C: Hazen-Williams C factor
   D: inside diameter of pipe, inch
   h: head loss due to friction, ft/1000 ft
Equation (4.27) can be transformed to solve for the head loss h in terms of flow rate Q and other variables as

$$h = 1.0461 \times 10^4 (Q/C)^{1.852} (1/D)^{4.87} \tag{4.27a}$$

It can be seen from Equations (4.27) and (4.27a) that as C increases, so does the flow rate. Also, the head loss h is approximately inversely proportional to the square of C and approximately directly proportional to the square of the flow rate Q.

In SI units, the Hazen-Williams formula is as follows:

$$Q = 9.0379 \times 10^{-8} (C)(D)^{2.63} (P_{km}/Sg)^{0.54} \tag{4.28}$$

$$h_{km} = 1.1323 \times 10^{12} (Q/C)^{1.852} (1/D)^{4.87} \tag{4.28a}$$

where

P_{km}: pressure drop due to friction, kPa/km.

h_{km}: pressure drop due to friction, m/km.

Q: flow rate, m^3/h

C: Hazen-Williams C factor

D: pipe inside diameter, mm

Sg: specific gravity of liquid, dimensionless

Appendix G lists the head loss in water pipelines using the Hazen-Williams equation with a C factor = 120 for various pipe sizes and flow rates in both USCS and SI units.

EXAMPLE 4.11 USCS UNITS

A 4-inch internal diameter smooth pipeline is used to pump 150 gal/min of water. Calculate the head loss in 3000 ft of pipe. Assume the value of Hazen-Williams C = 140.

Solution

Using Equation (4.27), the head loss h can be calculated as follows:

$$150 = 6.7547 \times 10^{-3} \, (140) \, (4.0)^{2.63} \, (h)^{0.54}$$

Solving for h, we get

$$h = 13.85 \text{ ft/1000 ft of pipe}$$

Therefore, the head loss for 3000 ft of pipe = $13.85 \times 3 = 41.55$ ft.

EXAMPLE 4.12 SI UNITS

A water pipeline has an internal diameter of 450 mm. Using a Hazen-Williams C factor = 120 and a flow rate of 500 m^3/h, calculate the head loss due to friction in m per km of pipe length.

Solution

Using Equation (4.28a), the head loss h can be calculated as follows:

$$h_{km} = 1.1323 \times 10^{12} (500/120)^{1.852} (1/450)^{4.87} = 1.91 \text{ m}$$

The head loss per km of pipe = 1.91 m.

EXAMPLE 4.13 SI UNITS

Based on measurements of pressures over the length of the pipeline, it is found that a water pipeline with an inside diameter of 480 mm has an average friction loss of 0.4 bar/km. The C factor is estimated to be in the range of 110–120 for this pipe. What is the flow rate for these conditions?

Solution

The given pressure loss due to friction is

$$P_{km} = 0.4 \text{ bar/km} = 0.4 \times 100 \text{ kPa/km} = 40 \text{ kPa/km}$$

Using Equation (4.28), the pressure drop is calculated as follows:

$$Q = 9.0379 \times 10^{-8} \ (C)(480)^{2.63}(40/1.0)^{0.54}$$

Solving for the flow rate Q, we get

For C = 110 $\quad$ Q = 820.8 m^3/h

For C = 120 $\quad$ Q = 895.4 m^3/h

For the given range of C values, the flow rate is between 820.8 and 895.4 m^3/h.

Pressure Loss through Fittings and Valves

Similar to friction loss through straight pipe, fittings and valves also cause pressure drop due to friction. These head losses in valves and fittings are collectively referred to as *minor losses* in a pipeline system. The reason for such categorization is because in most piping systems the actual head loss through straight pipe is several times that due to fittings and valves. However, in a terminal, tank farm, or a plant, there may be many short pieces of pipes and several fittings and valves. In comparison to straight lengths of pipe, these fittings and valves may contribute more than minor losses. On the other hand, in a long-distance pipeline of 10 miles or more, the contribution of the valves and fittings may be a small percentage of the total head loss, and thus they may be rightly called minor losses.

Although the mechanism is complicated, for turbulent flow we can simplify these minor losses by utilizing the concept of velocity head of the liquid. As explained in Chapter 1, the kinetic energy of flowing liquid is represented by the velocity head as follows:

$$\text{Velocity head} = V^2/2g \tag{4.29}$$

where

V: velocity of flow

g: acceleration due to gravity, 32.2 ft/s^2 (9.81 m/s^2)

Using USCS units, we see that the velocity head as the units of ft of liquid is as follows:

$$\text{Velocity head} = V^2/2g = (\text{ft/s})^2/(\text{ft/s}^2) = \text{ft}$$

Thus, velocity head represents the pressure expressed in ft of head of liquid. As an example, if V = 8 ft/s and g = 32.2 ft/s²:

$$\text{Velocity head} = 8 \times 8/(2 \times 32.2) = 0.9938 \text{ ft}$$

In SI units, for example, with V = 3 m/s and g = 9.81 m/s²:

$$\text{Velocity head} = 3 \times 3/(2 \times 9.81) = 0.4587 \text{ m}$$

The friction loss h_f in a valve or fitting is then defined as K times the velocity head just defined, where K is a dimensionless *resistance coefficient*. It is also sometimes referred to as the *head loss coefficient* or *energy loss coefficient* as follows:

$$h_f = K(V^2/2g) \tag{4.30}$$

The values of K for various valves are listed in Table 4.3.

For example, a 12-inch gate valve has a K value of 0.10. At a liquid velocity of 10 ft/s, the pressure drop through this 10-inch valve may be calculated using Equation (4.30) as follows:

$$h_f = 0.10 \times (10^2/64.4) = 0.1553 \text{ ft}$$

Table 4.3 for resistance coefficient K contains a column designated as L/D. This ratio is called the equivalent length ratio. The equivalent length is defined as a length of straight pipe that has the same head loss as the valve. For example, a fully open gate valve has an L/D ratio of 8, whereas a globe valve has an L/D ratio of 340. This means that a 16-inch gate valve is equivalent to 8 × 16 = 128 inches (or 10.7 ft) of straight pipe. Similarly, a 24-inch globe valve is equivalent to 340 × 24 = 8160 inches, or 680 ft, of straight pipe when calculating pressure drop.

For fittings such as elbows, tees, and so on, a similar approach can be used. Table 4.4 provides the K values and L/D ratios for commonly used fittings in the piping industry. The equivalent lengths of valves and fittings are tabulated in Table 4.5.

As an example, a DN 500 ball valve has L/D ratio = 3. Therefore, from a pressure drop standpoint, this valve is equivalent to a straight pipe of length 3 × 500 = 1500 mm, or 1.5 m. Using the concept of equivalent length, we can determine the total length of straight pipe in a piping system inclusive of all valves and fittings. Suppose on the discharge side of a pump, the straight pipe accounts for 5400 ft of NPS 16-inch pipe, and several valves and fittings are equivalent to a total

Table 4.3 Resistance Coefficient K for valves

| Description | L/D | Nominal Pipe Size – inches | | | | | | | | | | | |
		1/2	3/4	1	1-1/4	1-1/2	2	2-1/2 to 3	4	6	8 to 10	12 to 16	18 to 24
Gate Valve	8	0.22	0.20	0.18	0.18	0.15	0.15	0.14	0.14	0.12	0.11	0.10	0.10
Globe Valve	340	9.2	8.5	7.8	7.5	7.1	6.5	6.1	5.8	5.1	4.8	4.4	4.1
Ball Valve	3	0.08	0.08	0.07	0.07	0.06	0.06	0.05	0.05	0.05	0.04	0.04	0.04
Butterfly Valve							0.86	0.81	0.77	0.68	0.63	0.35	0.30
Plug Valve Straightway	18	0.49	0.45	0.41	0.40	0.38	0.34	0.32	0.31	0.27	0.25	0.23	0.22
Plug Valve 3-way thru-flo	30	0.81	0.75	0.69	0.66	0.63	0.57	0.54	0.51	0.45	0.42	0.39	0.36
Plug Valve branch - flo	90	2.43	2.25	2.07	1.98	1.89	1.71	1.62	1.53	1.35	1.26	1.17	1.08

Table 4.4 Resistance Coefficient K for Fittings

Description	L/D	Nominal Pipe Size – inches												
		1/2	3/4	1	1-1/4	1-1/2	2	2-1/2 to 3	4	6	8 to 10	12 to 16	18 to 24
Standard Elbow – 90°	30	0.81	0.75	0.69	0.66	0.63	0.57	0.54	0.51	0.45	0.42	0.39	0.36
Standard Elbow – 45°	16	0.43	0.40	0.37	0.35	0.34	0.30	0.29	0.27	0.24	0.22	0.21	0.19
Standard Elbow long radius 90°	16	0.43	0.40	0.37	0.35	0.34	0.30	0.29	0.27	0.24	0.22	0.21	0.19
Standard Tee thru-flo	20	0.54	0.50	0.46	0.44	0.42	0.38	0.36	0.34	0.30	0.28	0.26	0.24
Standard Tee thru-branch	60	1.62	1.50	1.38	1.32	1.26	1.14	1.08	1.02	0.90	0.84	0.78	0.72
Mitre bends - α = 0	2	0.05	0.05	0.05	0.04	0.04	0.04	0.04	0.03	0.03	0.03	0.03	0.02
Mitre bends - α = 30	8	0.22	0.20	0.18	0.18	0.17	0.15	0.14	0.14	0.12	0.11	0.10	0.10
Mitre bends - α = 60	25	0.68	0.63	0.58	0.55	0.53	0.48	0.45	0.43	0.38	0.35	0.33	0.30
Mitre bends - α = 90	60	1.62	1.50	1.38	1.32	1.26	1.14	1.08	1.02	0.90	0.84	0.78	0.72

Table 4.5 Equivalent lengths of valves and fittings

Description	L/D
Gate valve – fully open	8
Gate valve – ¾ open	35
Gate valve – ½ open	160
Gate valve – ¼ open	900
Globe valve – fully open	340
Angle valve – fully open	150
Ball valve	3
Butterfly valve – fully open	45
Check valve – swing type	100
Check valve – ball type	150
Standard Elbow – 90°	30
Standard Elbow – 45°	16
Long Radius Elbow–90°	16
Standard Tee – flow thru run	20
Standard Tee – flow thru branch	60

of 320 ft of 16-inch-diameter pipe Therefore, we can calculate the total head loss in the discharge piping by increasing the total length of pipe to 5400 + 320 = 5720 ft of pipe. If the pressure drop in a 16-inch-diameter pipe was calculated as 6.2 ft of liquid per 1000 ft of pipe, the total head loss due to friction in the 5720 ft of pipe is

$$H = 5720 \times 6.2/1000 = 35.46 \text{ ft of liquid}$$

Entrance and Exit Losses, and Losses Due to Enlargement and Contraction

In addition to the loss through fittings and valves, there are six other minor losses in piping systems:

1. Entrance loss
2. Exit loss
3. Loss due to sudden enlargement

4. Loss due to sudden contraction
5. Loss due to gradual enlargement
6. Loss due to gradual contraction

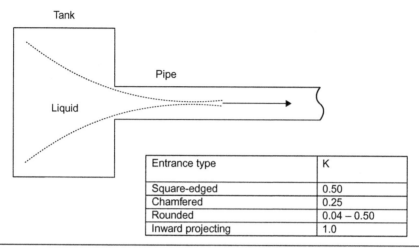

Entrance type	K
Square-edged	0.50
Chamfered	0.25
Rounded	0.04 – 0.50
Inward projecting	1.0

Figure 4.3   Entrance loss coefficient.

ENTRANCE LOSS

The entrance loss is the head loss that occurs when a liquid flows from a large tank into a pipe. At the entrance to the pipe, the liquid must accelerate from zero velocity at the liquid surface in the tank to the velocity corresponding to the flow rate through the pipe. This entrance loss can also be represented in terms of the velocity head as before.

$$\text{Entrance loss} = K(V^2/2g)$$

where K is the resistance coefficient or head loss coefficient that depends on the shape of the entrance as described in Figure 4.3. The value of K for entrance loss ranges from 0.04 to 1.0.

EXIT LOSS

The exit loss is associated with liquid flow from a pipe into a large tank as shown in Figure 4.4. As the liquid enters the tank, its velocity is decreased to very nearly zero. Similar to entrance loss, the exit loss can be calculated as

$$\text{Exit loss} = K(V^2/2g)$$

Generally, K = 1.0 is used for all types of pipe connection to a tank.

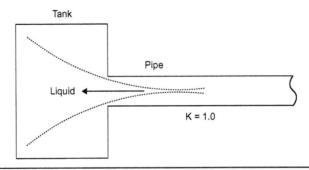

Figure 4.4 Exit loss.

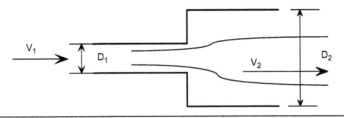

Figure 4.5 Sudden enlargement.

Loss Due to Sudden Enlargement

Sudden enlargement occurs when liquid flows from a smaller pipe to a larger pipe abruptly as shown in Figure 4.5. This causes the velocity to decrease abruptly, causing turbulence and, hence, energy loss. The head loss due to sudden enlargement can be represented by the head loss coefficient K, which depends on the ratio of the two pipe diameters. It can be calculated using the following equation:

$$K = [1 - (D_1/D_2)^2]^2 \qquad (4.31)$$

Loss Due to Sudden Contraction

Sudden contraction occurs when liquid flows from a larger pipe to a smaller pipe as shown in Figure 4.6. This causes the velocity to increase from the initial value of V_1 to a final value of V_2. Reviewing the flow pattern shows the formation of a throat, or *vena contracta*, immediately after the diameter change from D_1 to D_2, as shown in Figure 4.6. The value of resistance coefficient K for a sudden contraction is shown in Table 4.6 for various ratios A_2/A_1 of pipe cross-sectional areas. The ratio A_2/A_1 is easily calculated from the diameter ratio D_2/D_1 as follows:

$$A_2/A_1 = (D_2/D_1)^2 \qquad (4.32)$$

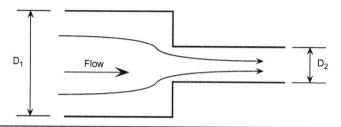

Figure 4.6 Sudden contraction.

Table 4.6 Head Loss Coefficient K for sudden contraction

A_2/A_1	D_2/D_1	K
0.0	0.0	0.50
0.1	0.32	0.46
0.2	0.45	0.41
0.3	0.55	0.36
0.4	0.63	0.30
0.5	0.71	0.24
0.6	0.77	0.18
0.7	0.84	0.12
0.8	0.89	0.06
0.9	0.95	0.02
1.0	1.0	0.00

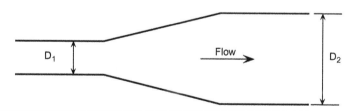

Figure 4.7 Gradual enlargement.

Loss Due to Gradual Enlargement

Gradual enlargement occurs when liquid flows from a smaller-diameter pipe to a larger-diameter pipe by a gradual increase in diameter, as depicted in Figure 4.7.

Table 4.7 Resistance coefficient for gradual enlargement

D_2/D_1	Cone Angle, θ												
	2°	6°	10°	15°	20°	25°	30°	35°	40°	45°	50°	60°
1.1	0.01	0.01	0.03	0.05	0.10	0.13	0.16	0.18	0.19	0.20	0.21	0.23
1.2	0.02	0.02	0.04	0.09	0.16	0.21	0.25	0.29	0.31	0.33	0.35	0.37
1.4	0.02	0.03	0.06	0.12	0.23	0.30	0.36	0.41	0.44	0.47	0.50	0.53
1.6	0.03	0.04	0.07	0.14	0.26	0.35	0.42	0.47	0.51	0.54	0.57	0.61
1.8	0.03	0.04	0.07	0.15	0.28	0.37	0.44	0.50	0.54	0.58	0.61	0.65
2.0	0.03	0.04	0.07	0.16	0.29	0.38	0.46	0.52	0.56	0.60	0.63	0.68
2.5	0.03	0.04	0.08	0.16	0.30	0.39	0.48	0.54	0.58	0.62	0.65	0.70
3.0	0.03	0.04	0.08	0.16	0.31	0.40	0.48	0.55	0.59	0.63	0.66	0.71

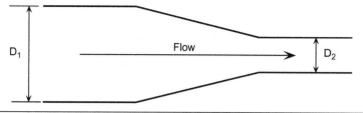

Figure 4.8 Gradual contraction.

The resistance coefficient K for a gradual enlargement depends on the angle of taper and the ratio of the two diameters. Table 4.7 provides values of K for various angles and diameter ratios. Generally, K values ranges from 0.01 to 0.72.

Loss Due to Gradual Contraction

Gradual contraction occurs when liquid flows from a larger-diameter pipe to a smaller-diameter pipe by a gradual decrease in the flow area, as depicted in Figure 4.8. Obviously, this is less severe than a sudden contraction, and the resistance coefficient for a gradual contraction will be less than that of a sudden contraction.

The resistance coefficient for a gradual contraction depends on the angle of the cone and ranges from 0.01 to a maximum of 0.36 for the cone angles up to 150 degrees. In most applications this coefficient is so small that it is generally neglected in comparison with other minor losses.

EXAMPLE 4.14 USCS UNITS

The suction piping to a pump consists of 53 ft of NPS 18 pipe, 0.250-inch wall thickness, two NPS 18 gate valves, and three NPS 18 LR elbows. The discharge piping from the pump to the delivery tank is composed of 9476 ft of NPS 16 pipe, 0.250-inch wall thickness, and the following valves and fittings: two NPS 16 gate valves (GV), one NPS 16 swing check valve, and six NPS 16 LR elbows. The meter manifold on the discharge side may be considered to be equivalent to a pressure drop of 10 psig. Determine the total pressure drop in the suction and discharge piping at a flow rate of 7800 gpm. The liquid specific gravity is 0.736, and viscosity is 0.6 cSt. Use the Hazen-Williams formula with a C factor = 140.

Solution

We will use the equivalent length method to convert all valves and fittings to straight pipe.

Suction piping:

2 – NPS 18 GV = 2 × 8 × 18/12 = 24 ft of NPS 18 pipe
3 – NPS 18 LR elbows = 3 × 16 × 18/12 = 72 ft of NPS 18 pipe
1 – straight pipe = 53 ft

The total equivalent length of straight pipe on suction side = 24 + 72 + 53 = 149 ft of NPS 18 pipe.

Discharge piping:

2 – NPS 16 GV = 2 × 8 × 16/12 = 21.33 ft of NPS 16 pipe
1 – NPS 16 swing check valve = 1 × 50 × 16/12 = 66.67 ft of NPS 16 pipe
6 – NPS 16 LR elbows = 6 × 16 × 16/12 = 128 ft of NPS 16 pipe
1 – straight pipe = 9476 ft

The total equivalent length of straight pipe on discharge side = 21.33 + 66.67 + 128 + 9476 = 9692 ft of NPS 16 pipe

Using Equation (4.27a), the head loss in NPS 18 pipe at 7800 gpm is calculated as follows:

$$h = 1.0461 \times 10^4 (7800/140)^{1.852} (1/17.5)^{4.87}$$

Solving for h, suction piping head loss h = 15.83 ft/1000 ft of pipe.

Similarly, for the discharge piping, the head loss in NPS 16 pipe at 7800 gpm is calculated as follows:

$$h = 1.0461 \times 10^4 (7800/140)^{1.852} (1/15.5)^{4.87}$$

Solving for h, discharge piping head loss h = 28.59 ft/1000 ft of pipe. Therefore, the total pressure drop in the suction piping is

$$\Delta P = 15.83 \times 149/1000 = 2.36 \text{ ft of liquid}$$
$$= 2.36 \times 0.736/2.31$$
$$\Delta P = 0.752 \text{ psig}$$

Similarly, the total pressure drop in the discharge piping is

$$\Delta P = 28.59 \times 9692/1000 = 277.09 \text{ ft of liquid}$$
$$= 277.09 \times 0.736/2.31$$
$$= 88.29 \text{ psig} + 10 \text{ psig for meter manifold}$$
$$\Delta P = 98.29 \text{ psig}$$

EXAMPLE 4.15 SI UNITS

The suction piping to a pump consists of 32 m of 500 mm outside diameter pipe, 10 mm wall thickness, two DN 500 gate valves, and three DN 500 LR elbows. The discharge piping from the pump to the delivery tank is composed of 3150 m of 400 mm outside diameter pipe, 8 mm wall thickness, and the following valves and fittings: two DN 400 gate valves (GV), one DN 400 swing check valve, and six DN 400 LR elbows. The meter manifold on the discharge side may be considered to be equivalent to a pressure drop of 1.2 bar. For a flow rate of 31,200 L/min of water, using Hazen-Williams formula and C factor = 120, calculate the total pressure drop in the suction and discharge piping.

Solution

We will use the equivalent length method to convert all valves and fittings to straight pipe.

Suction piping:

2 – DN 500 GV = 2 × 8 × 500/1000 = 8 m of DN 500 pipe
3 – DN 500 LR elbows = 3 × 16 × 500/1000 = 24 m of DN 500 pipe
1 – straight pipe = 32 m DN 500 pipe

The total equivalent length of straight pipe = 8 + 24 + 32 = 64 m DN 500.

Discharge piping:

2 – DN 400 GV = 2 × 8 × 0.4 = 6.4 m of DN 400 pipe
1 – DN 400 swing check valve = 1 × 50 × 0.4 = 20 m of DN 400 pipe
6 – DN 400 LR elbows = 6 × 16 × 0.4 = 38.4 m of DN 400 pipe
1 – straight pipe = 3150 m DN 400 pipe

The total equivalent length of straight pipe = 6.4 + 20 + 38.4 + 3150 = 3214.8 m of DN 400 pipe.

Using Equation (4.28), the pressure loss in the 500 mm outside diameter pipe at 31,200 L/min is calculated as follows:

$$31200 \times 24/1000 = 9.0379 \times 10^{-8}(120)(480)^{2.63}(P_{km}/1.0)^{0.54}$$

Solving for P_{km}, we get:

Suction piping pressure loss P_{km} = 28.72 kPa/km of pipe length

Similarly for the discharge piping, the pressure loss in 400 mm outside diameter pipe at 31,200 L/min is calculated as follows"

$$31200 \times 24/1000 = 9.0379 \times 10^{-8}(120)(384)^{2.63}(P_{km}/1.0)^{0.54}$$

Solving for P_{km}, we get:

Discharge piping pressure loss P_{km} = 85.16 kPa/km of pipe

Therefore, the total pressure drop in the suction piping is

$$\Delta P = 28.72 \times 64/1000 = 1.84 \text{ kPa}$$

Similarly, the total pressure drop in the discharge piping is

$$\Delta P = 85.16 \times 3214.8/1000 = 273.77 \text{ kPa}$$
$$= 273.77 \text{ kPa} + 1.2 \text{ bar for meter manifold}$$
$$\Delta P = 393.77 \text{ kPa or } 3.94 \text{ bar}$$

EXAMPLE 4.16 USCS UNITS

An example of a gradual enlargement is an NPS 10 pipe to an NPS 12 pipe with an included angle of 30 degrees. If the nominal wall thickness is 0.250 inch, calculate the head loss due to the gradual enlargement at a flow rate of 2000 gpm of water.

Solution

The velocity of water in the two pipe sizes is calculated using Equation (4.4):

$$V_1 = 0.4085 \times 2000/(10.75 - 0.25 \times 2)^2 = 7.77 \text{ ft/s}$$

$$V_2 = 0.4085 \times 2000/(12.75 - 0.25 \times 2)^2 = 5.44 \text{ ft/s}$$

Diameter ratio D_2/D_1 = 12.25/10.25 = 1.195

From Table 4.7, for D_2/D_1 = 1.195 and cone angle = 30 degrees, K = 0.25. Head loss using Equation (4.30) is

$$h_f = 0.26 \times (7.77)^2/64.4 = 0.2437 \text{ ft}$$

EXAMPLE 4.17 SI UNITS

A gradual contraction occurs from 400 to 500 mm outside diameter pipe with a nominal wall thickness of 8 mm. What is the head loss when water flows through this piping system at 32,000 L/min, assuming K = 0.03?

Solution

The velocity in the smaller pipe is calculated using Equation (4.6):

$$V = 353.6777(32 \times 60)/(384)^2 = 4.605 \text{ m/s}$$

The head loss is calculated using Equation (4.30) as

$$h_f = 0.03 \times (4.605)^2/(2 \times 9.81) = 0.0324 \text{ m}$$

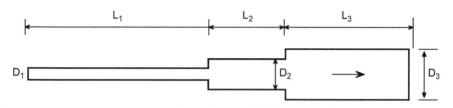

Figure 4.9 Series piping.

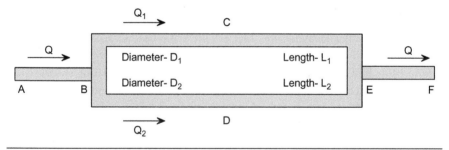

Figure 4.10 Parallel piping.

Pipes in Series and Parallel

In the preceding pages we have reviewed head loss in piping systems consisting of straight lengths of pipe, valves, and fittings. Frequently, pump systems and pump stations include different pipe sizes that are arranged in series and parallel configuration. An example of a series piping system is shown in Figure 4.9. Figure 4.10 shows an example of a parallel piping system consisting of two pipes connected in a loop joined to a single pipe at the beginning and end of the loop.

When two or more pipe segments are connected such that the liquid flows from one pipe segment to another in succession, the arrangement is called series piping.

The flow rate through each pipe segment is the same, and the pressure drops in each segment are added together to get the total head loss in the piping system.

In parallel piping, two or more pipe sections are connected such that starting with a single pipe at the entry, the flow is split into the multiple pipes and they rejoin at the downstream end into a single pipe as shown in Figure 4.10. We will review how the frictional head loss is calculated in series and parallel piping in the following pages.

Head Loss in Series Piping

One approach to calculate the head loss in series piping is to consider each pipe segment separately and determine the head loss for the common flow rate and add the individual pressure drops to obtain the total head loss in the series piping system. Thus, referring to Figure 4.9, we would calculate the head loss in the pipe of diameter D_1 and length L_1, followed by the head loss for diameter D_2 and length L_2, and so on for all pipe segments at the common flow rate Q. The total head loss for this series arrangement is then $\Delta P = \Delta P_1 + \Delta P_2 + \Delta P_3$. This will be illustrated using an example.

Another approach to calculating the total head loss in series piping is to consider one of the given pipe diameters as the base diameter and convert the other pipe sizes to equivalent lengths of the base diameter. The equivalent length is defined as the length of a particular diameter pipe that has the same head loss. Suppose we have 2500 ft of 14-inch pipe and 1200 ft of 16-inch pipe connected in the series. We will use the 14-inch pipe as the base or reference diameter. The 1200 ft of 16-inch pipe will be converted to some equivalent length L of the 14-inch pipe such that the head loss in the 1200 ft of 16-inch pipe is equal to the head loss in L ft of 14-inch pipe. Now the series piping can be considered to consist of a single piece of 14-inch-diameter pipe, with an equivalent total length of (2500 + L). Once we calculate the pressure drop in the 14-inch pipe (in ft/1000 ft or psi/mi), we can calculate the total head loss by multiplying the unit head loss by the total equivalent length.

Similarly, for more than two pipes in series, the total equivalent length of pipe with the base diameter will be the sum of the individual equivalent lengths. The head loss can then be calculated for the base pipe diameter and the total head loss determined by multiplying the head loss per unit length by the total equivalent length.

Consider a pipe A with an inside diameter D_A and length L_A connected in series with another pipe B with an inside diameter D_B and length L_B. We can replace this two-pipe system with a single pipe with an equivalent diameter D_E and length L_E such that the two piping systems have the same pressure drop at the given flow rate. The equivalent diameter D_E may be chosen as one of the given diameters.

From previous discussions and Equation (4.17) we know that the pressure drop due to friction is inversely proportional to the fifth power of the diameter and directly proportional to the pipe length. Setting the base diameter as D_A, the total equivalent length L_E of the two pipes in terms of pipe diameter D_A is

$$L_E/(D_A)^5 = L_A/(D_A)^5 + L_B/(D_B)^5 \qquad (4.33)$$

Simplifying, we get the following for the total equivalent length:

$$L_E = L_A + L_B(D_A/D_B)^5 \qquad (4.34)$$

Using the equivalent length L_E based on diameter D_A, the total pressure drop in length L_E of pipe diameter D_A will result in the same pressure drop as the two pipe segments of lengths L_A and L_B in series. Note that this equivalent length approach is only approximate. If elevation changes are involved, it becomes more difficult to take those into account.

EXAMPLE 4.18 USCS UNITS

Consider a series piping system consisting of an NPS 16 pipe, 0.250 inch wall thickness, and 6850 ft long and an NPS 18 pipe, 0.250 inch wall thickness, and 11,560 ft long. Calculate the equivalent length in terms of NPS 16 pipe and the total head loss due to friction at a flow rate of 5000 gpm of water. Use the Hazen-Williams equation with C = 120. Ignore elevation differences.

Solution

The equivalent length from Equation (4.34) is calculated as

$$L_E = 6850 + 11560((16 - 2 \times 0.25)/(18 - 2 \times 0.5))^5$$
$$= 6850 + 6301.24 = 13{,}151.24 \text{ ft}$$

The head loss due to friction based on the NPS 16 pipe is from Equation (4.27a):

$$h = 1.0461 \times 10^4 (5000/120)^{1.852}(1/15.5)^{4.87}$$

Solving for the head loss h in ft/1000 ft of pipe:

$$h = 16.69 \text{ ft/1000 ft}$$

The total head loss $= 16.69 \times 13151.24/1000 = 219.5$ ft.

EXAMPLE 4.19 SI UNITS

A series piping system consists of a 500 mm outside diameter pipe, 10 mm wall thickness, and 1200 m long and a DN 400 pipe, 8 mm wall thickness and 940 m long. Calculate the equivalent length based on the DN 400 pipe and the total head loss due to friction at a flow rate of 800 m³/h of water. Use the Hazen-Williams equation with C = 110. Ignore elevation differences.

Solution

Using Equation (4.34), the equivalent length based on DN 400 pipe is

$$L_E = 940 + 1200((400 - 2 \times 8)/(500 - 2 \times 10))^5$$
$$= 940 + 393.22 = 1333.22 \text{ m}$$

The head loss using the Hazen-Williams Equation (4.28) is calculated as follows:

$$800 = 9.0379 \times 10^{-8} \, (110) \, (400 - 2 \times 8)^{2.63} \, (Pkm)^{0.54}$$

Solving for head loss in kPa/km

$$P_{km} = 113.09 \text{ kPa/km}$$

The total head loss = 113.09 × 1333.22/1000 = 150.75 kPa = 1.508 bar.

Head Loss in Parallel Piping

When two pipes are connected in parallel, the individual flow rate through each pipe is calculated by considering a common pressure drop between the inlet and the outlet of the pipe loop. In the Figure 4.10 pipe segment, AB contains the inlet flow rate Q, which splits into Q_1 in pipe segment BCE and Q_2 in BDE. The flows then rejoin at E to form the outlet flow Q, which then flows through the pipe segment EF. Since pipe segments BCE and BDE are in parallel, the common junction pressures at B and E causes identical pressure drops in BCE and BDE. This common pressure drop in the two parallel pipe segments is used to determine the flow split in the segments. Using the Darcy equation (4.11) and subscript 1 for pipe segment BCE and subscript 2 for the pipe segment BDE, the pressure drops can be stated as follows:

$$h_1 = f_1 (L_1/D_1) V_1^2/2g \qquad (4.35)$$

$$h_2 = f_2 (L_2/D_2) V_2^2/2g \qquad (4.36)$$

where V_1 and V_2 are the flow velocities in the pipe segments BCE and BDE, and f_1 and f_2 are the corresponding friction factors. If we assume initially that $f_1 = f_2 = f$, a common friction factor, equating the two head losses and simplifying we get

$$(L_1/D_1)V_1^2 = (L_2/D_2)V_2^2 \tag{4.37}$$

The velocities can be represented in terms of the flow rates Q_1 and Q_2 using Equation (4.4) as follows:

$$V_1 = 0.4085\ Q_1/D_1^2$$
$$V_2 = 0.4085\ Q_2/D_2^2$$

Substituting these in Equation (4.37) and simplifying, we get the following relationship between Q_1 and Q_2:

$$Q_1/Q_2 = (L_2/L_1)^{0.5}(D_1/D_2)^{2.5} \tag{4.38}$$

If the friction factors f_1 and f_2 are not assumed to be the same, Equation (4.38) becomes

$$Q_1/Q_2 = (f_2 L_2/f_1 L_1)^{0.5}(D_1/D_2)^{2.5} \tag{4.38a}$$

Since the sum of Q_1 and Q_2 must equal the inlet flow Q, we have

$$Q_1 + Q_2 = Q \tag{4.39}$$

From Equations (4.38) and (4.39), we can calculate the two flow rates Q_1 and Q_2 in terms of the inlet flow Q. In this analysis, we assumed the value of the friction factor f was the same in both pipe segments. This is a fairly good approximation for a start. Once we get the values of the individual flow rates Q_1 and Q_2 by solving Equations (4.38) and (4.39), we can then calculate the Reynolds numbers for each pipe segment and calculate the value of f using the Colebrook-White equation or the Moody diagram. Using these new values of f, the head losses can be found from Equations (4.35) and (4.36) and the process repeated to calculate Q_1 and Q_2. The iteration is continued until the successive values of the flow rates are within some tolerance—say, 1% or less.

Another approach to calculating the head loss in parallel pipes is to determine the *equivalent diameter*. The two parallel pipes may be replaced by a single pipe with an equivalent diameter D_E and length L_E. Since this equivalent single pipe replaces the two parallel pipes, it must have the same pressure drop between points B and E. Therefore, using Equation (4.37) we can state that

$$(L_E/D_E)V_E^2 = (L_1/D_1)V_1^2 \tag{4.40}$$

where V_E is the flow velocity in the equivalent diameter D_E, which handles the full flow Q. Since velocity is proportional to flow rate, V_E can be replaced with the equivalent flow rate Q and V_1 replaced with Q_1 so that Equation (4.40) simplified as

$$(L_E/D_E)(Q/D_E^2)^2 = (L_1/D_1)(Q_1/D_1^2)^2 \qquad (4.41)$$

As before, if the friction factors f_1 and f_2 are not assumed to be the same, Equation (4.41) becomes

$$(f_E L_E/D_E)(Q/D_E^2)^2 = (f_1 L_1/D_1)(Q_1/D_1^2)^2 \qquad (4.41a)$$

Setting $L_E = L_1$ and simplifying further, we get the following relationship for the equivalent diameter D_E for the single pipe that will replace the two parallel pipes with the same head loss between points B and E:

$$D_E = D_1(f_E/f_1)^{0.2}(Q/Q_1)^{0.4} \qquad (4.42)$$

With the simplifying assumption of $f_1 = f_E$, this equation becomes

$$D_E = D_1(Q/Q_1)^{0.4} \qquad (4.42a)$$

Equation (4.42) can be applied to the second parallel pipe segment, and we get the following relationship:

$$D_E = D_2(f_E L_E/f_2 L_2)^{0.2}(Q/Q_2)^{0.4} \qquad (4.43)$$

Again, with the assumption of $f_E = f_2$ and $L_E = L_1$, this equation becomes

$$D_E = D_2(L_1/L_2)^{0.2}(Q/Q_2)^{0.4} \qquad (4.43a)$$

Using Equations (4.39), (4.42), and (4.43) we can solve for the three unknowns D_E, Q_1, and Q_2 either assuming common friction or different friction factors (more accurate). Of course, the latter would require iterative solution, which would take a little longer.

The equivalent diameter approach is useful when you have to calculate the head loss through a piping system for several different inlet flow rates, especially when creating a system head curve. In the next example, the equivalent diameter concept will be examined to illustrate calculation of head loss in a parallel piping system.

EXAMPLE 4.20 USCS UNITS

An NPS 12 water pipeline from A to B is 1500 ft long. At B, two parallel pipe segments, each 4200 ft long, NPS 10 and NPS 12, are connected that rejoin at point D. From D, a single NPS 12 pipe, 2800 ft long extends to point E as shown in Figure 4.10. All pipes are 0.250 inch wall thickness. Assuming that the friction factors are the same for the parallel pipe segments, determine the flow split between the two parallel pipes for an inlet flow of 3200 gpm. Calculate the pressure drops in the three-pipe segment AB, BD, and DE. Compare results using the equivalent diameter method. Use the Hazen-Williams equation, C = 120.

Solution

The flow split between the two parallel pipes will be calculated first. Using Equations (4.38) and (4.39)

$$Q_1 + Q_2 = 3200$$

And

$$Q_1/Q_2 = (4200/4200)^{0.5}((10.75 - 2 \times 0.25)/(12.75 - 2 \times 0.25))^{2.5} = 0.6404$$

Substituting in the flow rate equation, replacing Q_1 with $0.6404 \times Q_2$

$$0.6404 \times Q_2 + Q_2 = 3200$$

Therefore, $Q_2 = 3200/1.6404 = 1950.74$ gpm, the flow rate in the NPS 12 pipe. The flow in the NPS 10 pipe is

$$Q_1 = 3200 - 1950.74 = 1249.26 \text{ gpm}$$

Using the Hazen-Williams equation (4.27a), the head loss in the three pipe segments can be calculated as follows:

For AB: $h_{AB} = 1.0461 \times 10^4 (3200/120)^{1.852} (1/12.25)^{4.87}$
For BD (NPS10): $h_{BD} = 1.0461 \times 10^4 (1249.26/120)^{1.852} (1/10.25)^{4.87}$
For DE: $h_{DE} = 1.0461 \times 10^4 (3200/120)^{1.852} (1/12.25)^{4.87}$

Solving for the head loss:

$h_{AB} = 22.97$ ft/1000 ft of pipe
$h_{BD} = 9.59$ ft/1000 ft of pipe
$h_{DE} = 22.97$ ft/1000 ft of pipe

The total head losses in the three pipe segments are as follows:

AB: Total head loss = 22.97 × 1500/1000 = 34.46 ft
BD: Total head loss = 9.59 × 4200/1000 = 40.28 ft
DE: Total head loss = 22.97 × 1500/1000 = 34.46 ft

Therefore, the total head loss from A to E = 34.46 + 40.28 + 34.46 = 109.20 ft. The equivalent diameter for the parallel pipes is calculated using Equations (4.39), (4.42), and (4.43) and assuming common friction factors, $Q_1 + Q_2 = Q$ from Equation (4.39), $D_E = D_1 (Q/Q_1)^{0.4}$ from Equation (4.42a), and $D_E = D_2 (L_1/L_2)^{0.2} (Q/Q_2)^{0.4}$ from Equation (4.43a).

Equating the two values of D_E, we get the following:

$$D_E = 10.25(3200/Q_1)^{0.4} = 12.25(4200/4200)^{0.2}(3200/Q_2)^{0.4} \qquad (4.44)$$

$$10.25/12.25 = (Q_1/Q_2)^{0.4}, \text{ or}$$

$$Q_1 = 0.6404 \times Q_2$$

Therefore, $0.6404 \times Q_2 + Q_2 = 3200$, and $Q_2 = 3200/1.6404 = 1950.74$ gpm, the flow rate in the NPS 12 pipe.

The equivalent diameter is then found from Equation (4.44) as

$$D_E = 12.25(4200/4200)^{0.2}(3200/1950.74)^{0.4} = 14.93 \text{ in.}$$

Thus, the equivalent diameter pipe must have an inside diameter of 14.93 in.

The pressure drop in BD using the equivalent diameter is as follows:

$$h_{BD} = 1.0461 \times 10^4 (3200/120)^{1.852} (1/14.93)^{4.87}, \text{ or}$$
$$h_{BD} = 8.77 \text{ ft/1000 ft of pipe}$$

Compare this with $h_{BD} = 9.59$ that we calculated before. The difference is (9.59 − 8.77)/9.59 = 0.0855, or 8.55%. If we had not assumed the same friction factors, we would have to calculate the Reynolds number for the first approximation flow rates $Q_1 = 1249.26$ and $Q_2 = 1950.74$ and then find the friction factor for each pipe using the Moody diagram. Next we use Equations (4.38a) and (4.39) to solve for Q_1 and Q_2 the second time. The process is repeated until the successive values of the flow rates are within some tolerance—say, 1% or less.

EXAMPLE 4.21 SI UNITS

A parallel piping system, similar to Figure 4.10, consists of three sections AB, BD, and DE. AB and DE are DN 500 with 10 mm wall thickness. The section BD consists of two parallel pipes, each DN 300 with 8 mm wall thickness. The pipe lengths are AB = 380 m, BD = 490 m, and DE = 720 m. Water enters point A at a flow rate of 950 m^3/h and flows through the parallel pipes to delivery point E. Determine the pressure drops in each segment and the total head loss between A and E. What diameter single pipe 490 m long can replace the two parallel pipes without changing the total pressure drop? Use the Hazen-Williams equation, with C = 110.

Solution

We will first determine the flow split in the two parallel pipes. Using Equations (4.38) and (4.39), the flows Q_1 and Q_2 are as follows:

$$Q_1/Q_2 = (490/490)^{0.5}((300 - 16)/(300 - 16))^{2.5} = 1$$

$$Q_1 + Q_2 = 950, \text{ or } Q_1 = Q_2 = 475 \text{ m}^3/\text{h}$$

This could have been inferred, since both parallel pipes are identical in size, and thus each pipe should carry half the inlet flow.

The head loss using the Hazen-Williams Equation (4.28) is calculated for each segment as follows:

For AB: $950 = 9.0379 \times 10^{-8} (110)(500 - 2 \times 10)^{2.63} (P_{AB})^{0.54}$
For BD: $475 = 9.0379 \times 10^{-8} (110)(300 - 2 \times 8)^{2.63} (P_{BD})^{0.54}$
For DE: $950 = 9.0379 \times 10^{-8} (110)(480)^{2.63} (P_{DE})^{0.54}$

Since AB and DE are both DN 500 pipes, the pressure drop per unit length will be the same.

For AB or DE, we get by simplifying

$$P_{AB} = P_{DE} = 52.45 \text{ kPa/km}$$

And

$$P_{BD} = 187.25 \text{ kPa/km}$$

Therefore, the head losses in the three segments are as follows:

Head loss in AB = $52.45 \times 0.950 = 49.83$ kPa
Head loss in BD = $187.25 \times 0.490 = 91.75$ kPa
Head loss in DE = $52.45 \times 0.720 = 37.76$ kPa

The total head loss between A and E = 49.83 + 91.75 + 37.76 = 179.34 kPa.

The equivalent diameter pipe to replace the two parallel pipes is calculated using Equation (4.42):

$$D_E = 284(950/475)^{0.4} = 374.74 \text{ mm}$$

Thus, the equivalent diameter pipe should have an inside diameter of approximately 375 mm.

Summary

In this chapter, we discussed the calculation methodology for pressure loss when a liquid flows through a pipeline. The calculation of velocity of flow, the Reynolds number, and the different types of flow were reviewed. The Darcy equation for head loss was introduced, and the dependence of the friction factor on the Reynolds number was explained. Determining the friction factor using the Moody diagram and the alternative method of using the Colebrook-White equation were illustrated by example problems. The importance of the Hazen-Williams equation in the water pipeline industry was discussed, and several example cases using the Hazen-Williams equation for calculating flow rate from pressure drop and vice versa were shown.

The minor losses associated with fittings and valves were explained using the resistance coefficient and velocity head. Alternatively, the concept of using the equivalent length to calculate minor losses was illustrated. The calculation of the total pressure drop in a piping system consisting of pipe, fittings, and valves was illustrated using examples. Other minor losses such as entrance loss, exit loss, and losses due to enlargement and contraction of pipes were also discussed. Series and parallel piping systems and the calculation of the head losses using the equivalent length and equivalent diameter method were also explained. In the next chapter we will discuss the development of system head curves and determining the operating point on a pump curve.

Problems

4.1 Water flows through a DN 400 pipe, 8 mm wall thickness at 1000 m³/h. Calculate the average flow velocity.

4.2 Calculate the Reynolds number in a water pipeline NPS 18, 0.281 in. wall thickness for a flow rate of 4200 bbl/h. Sg = 1.0, viscosity = 1.0 cSt.

4.3 Determine the minimum flow rate required to maintain turbulent flow in a gasoline pipeline (Sg = 0.74 and viscosity = 0.6 cSt at 20°C) if the pipe is DN 300, 6 mm wall thickness.

4.4 Turbine fuel flows at 3200 gpm in an NPS 14, 0.250 in. wall thickness pipeline. Calculate the Reynolds number and the head loss due to friction using the Darcy equation, assuming friction factor f = 0.015. The properties of turbine fuel at flowing temperature are Sg = 0.804 and viscosity = 1.92 cP.

4.5 A crude oil pipeline, DN 500, 8 mm wall thickness transports the product at a flow rate of 1325 m³/h. Use the Moody diagram to calculate the friction factor and the head loss per unit length of pipeline. The crude oil has the following properties at pumping temperature: Sg = 0.865 and viscosity = 17.2 cSt.

4.6 A pipeline 50 in. inside diameter is used to pump water at a flow rate of 20,000 gpm. Determine the head loss using the Hazen-Williams equation and C = 120.

4.7 The suction piping between a gasoline storage tank and the pump consists of 29 m of DN 400, 8 mm wall thickness pipe in addition to the following valves and fittings: two DN 400 gate valves and three DN 400 LR elbows. Calculate the total equivalent length of straight pipe. What is the head loss in the suction piping at a pumping rate of 3200 L/min?

4.8 A series piping system consists of three pipes as follows:
3000 ft of NPS 16 pipe, 0.281 in. wall thickness.
6150 ft of NPS 14 pipe, 0.250 in. wall thickness
1560 ft of NPS 12 pipe, 0.250 in. wall thickness
Calculate the equivalent length in terms of NPS 16 pipe and the total head loss due to friction at a flow rate of 4250 gpm of water. Use the Hazen-Williams equation with C = 110. Ignore elevation differences.

4.9 A water pipeline consists of two parallel pipes DN 300, 6 mm wall thickness, each 650 m long connected to a DN 400, 8 mm wall thickness, 1200 m long, similar to Figure 4.10. Pipe EF is DN 500, 8 mm wall thickness, and 1850 m long. Determine the flow split between the two parallel pipes for an inlet flow of 2900 L/min. Calculate the head loss in all pipe segments. Compare results using the equivalent diameter method. Use the Hazen-Williams equation, C = 110.

Chapter 5

System Head Curves

In Chapter 3, we discussed the method of determining the head required of a pump for a specific application. We found that in order to transport diesel fuel at a flow rate of 2800 gal/min from Hartford to Compton, a pump must be installed at Hartford that would generate a differential head of 510 ft at a capacity of 2800 gal/min. This was based on calculating the total pressure loss due to friction in the piping system between Hartford and Compton.

If we were interested in pumping 1500 gal/min instead, we would have calculated the head requirement as some number lower than 510 ft. Conversely, if we desired to pump at a faster rate, such as 4000 gal/min, we would have calculated the head required at a value higher than 510 ft. This is because the head required is directly

DOI: 10.1016/B978-1-85617-828-0.00005-6

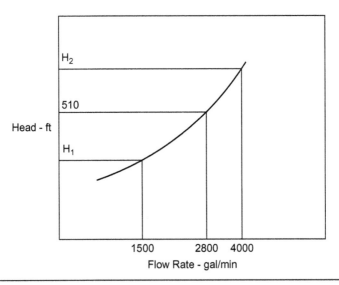

Figure 5.1 System head curve.

proportional to the friction loss in the piping, which in turn increases with increase in flow rate. Thus, we conclude that the head required to pump diesel fuel at various flow rates from Hartford to Compton would increase as the flow rate increases, as shown in Figure 5.1.

In Figure 5.1, the flow rate is plotted on the horizontal axis, while the vertical axis shows the head required (H). The resulting curve that shows the variation of head (H) with flow rate (Q) is called the *system head curve* and is characteristic of the piping system between the origin and the destination. Sometimes the term *system curve* is also used. The general shape of the system head curve is a parabola or second-degree equation in the flow rate Q. The reason for the concave shape (as opposed to a convex shape of the pump head curve) is because as the flow rate Q increases along the horizontal axis, the value of the system head increases approximately as the second power of Q. In other words, the head H varies as Q^2. As we observed in Chapter 4, the head loss is proportional to the square of the flow rate in the Darcy equation, whereas in the Hazen-Williams equation, the head loss varies as $Q^{1.852}$. Thus, similar to our analysis of pump head curves in Chapter 2, we can express H as a function of Q as follows:

$$H = a_0 + a_1 Q + a_2 Q^2,$$

where the constants a_0, a_1, and a_2 depend on the piping system geometry, such as length, diameter, and so on.

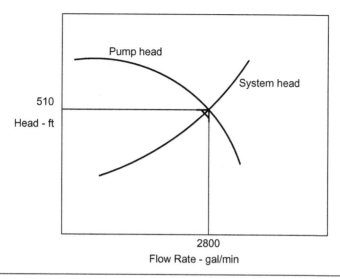

Figure 5.2 Pump head curve – System head curve.

Notice in Figure 5.1 that the point that represents 2800 gpm on the horizontal axis and 510 ft of head on the vertical axis was the operating condition for the pump selection in Example 3.6. The pump that we selected for this application exactly matches the condition Q = 2800 gal/min, H = 510 ft. We could then plot the H-Q curve of the pump superimposed on the system head curve as shown in Figure 5.2.

The point of intersection between the system head curve and the pump head curve represents the operating point (or duty point) of the pump. If we had chosen a larger pump, its head curve would be located above the present curve, resulting in an operating point that has larger Q and H values. Conversely, if a smaller pump were selected, the corresponding operating point would be at a lower flow rate and head. Thus, we conclude that when a pump head curve is superimposed on a system head curve, the flow rate and head corresponding to the point of intersection represent the operating point of the pump curve. It represents the pump capacity at which the head developed by the pump exactly matches the head required by the pipeline system, no more or no less. In summary, if we plotted a system head curve for a pipeline, we can determine the operating condition for a particular pump curve from the point of intersection between the two head curves.

The development of the system head curve for any piping system is the first task to be performed to determine the flow rate possible using a particular pump. We will illustrate the development of the system head curve using the data from Example 3.6. We already have the head calculations for Q = 2800 gpm: H = 510 ft.

Similar calculations will be performed for two additional flow rates: Q = 1000 gpm and Q = 3500 gpm. For each of these flow rates, we will calculate the pressure drop in the 16-inch suction piping and the 12-inch discharge piping, using the method outlined in Chapter 4. Once the pressure drop is known, the total head required at these two flow rates can be calculated, as we did before for Q = 2800 gpm.

For Q = 1000 gpm, the following pressure drops are calculated using the method outlined in Chapter 4, assuming an absolute roughness of 0.002 in for each pipe. First the Reynolds number was calculated, and then the friction factor f was found using the Moody diagram or the Colebrook-White equation. Using f and the pipe diameter, liquid specific gravity, and the flow rate, the pressure drop is found using Equation (4.12):

Pressure drop for 16-inch pipe = 1.51 psi/mi
Pressure drop for 12-inch pipe = 4.68 psi/mi

Please refer to Example 3.6 for details of these calculations.

The total pressure drop of the discharge piping 5.54 mi long, including the pressure drop through the two meter manifolds is

$$5.54 \times 4.68 + 15 + 15 = 55.93 \text{ psi}$$

Minimum discharge pressure required of the pump is

$$55.93 + (40 \times 0.85 / 2.31) = 70.65 \text{ psi (approximately)}$$

Pump differential pressure required is

$$70.65 - (25.76 - 1.51 \times 30 / 5280) = 44.9 \text{ psi}$$

Converting to ft of head, we get

$$44.9 \times 2.31 / 0.85 = 122 \text{ ft}$$

Similarly, for Q = 3500 gpm, the following pressure drops are calculated using the same approach as for Q = 1000 gpm:

Pressure drop for 16-inch pipe = 14.52 psi/mi
Pressure drop for 12-inch pipe = 45.94 psi/mi

Pump differential pressure required is

$$5.54 \times 45.94 + 30 + (40 \times 0.85/2.31) - (25.76 - 14.52 \times 30/5280) = 274 \text{ psi}$$

Converting to ft of head, we get

$$274 \times 2.31 / 0.85 = 745 \text{ ft}$$

The following table presents the Q, H values for the system head curve. The H-Q values for the system curve are plotted in Figure 5.3.

System Head Curve

Q-gpm	1000	2800	3500
H-ft	122	510	745

As mentioned earlier, the shape of the system head curve is approximately a parabola. The steepness of the curve is a function of the pressure drop in the pipeline. Since the pipeline is 12 inches in diameter, if we increase the pipe size to 16 inches in diameter, this will cause a reduction in pressure drop. This in turn will cause the system curve to flatten and move to the right. If the pipe were 10 inches instead of 12 inches, the pressure drop would increase and the system head curve would be steeper and move to the left, as shown in Figure 5.4.

Similarly, keeping the discharge pipe size the same (12 inches), if we change the product from diesel to gasoline, the pressure drop will decrease (due to lower specific gravity and viscosity of gasoline) and the system head curve will be flatter and to the right, as shown in Figure 5.5.

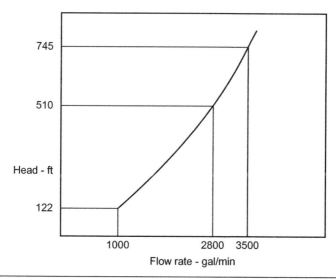

Figure 5.3 System head curve development.

Superimposing the pump head curve on the diesel and gasoline system curves, we note that the operating point moves from point A for diesel to point B for gasoline. It is clear that a higher flow rate is achieved when pumping gasoline compared to when pumping diesel, due to the lower specific gravity and viscosity of gasoline. Suppose we selected a pump based on pumping gasoline at the flow rate corresponding to point B in Figure 5.5. Later, we decide to pump diesel with this same pump. The diesel flow rate that can be achieved is represented by point A, which is lower than that at point B, with gasoline. Therefore, it is important to remember that when selecting a pump for multiproduct pumping, if the original pump selection was based on the heavier, more viscous product, switching to the lighter fluid will increase the pipeline throughput. Also, since the pump power requirement changes

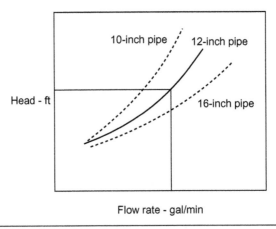

Figure 5.4   System head curves – Different pipe sizes.

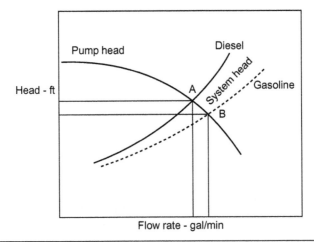

Figure 5.5   System head curves – Pumping different liquids.

with the product, the installed motor driver must be large enough for the worst-case scenario. The lower flow rate and higher specific gravity of diesel may demand more pump power than the higher flow rate and lower specific gravity gasoline.

Pump Throttling and Power Loss

There are times when the pump flow rate may have to be reduced by partially closing a valve on the discharge side of the pump. This may be necessary to prevent pump cavitation at higher flow rates or to reduce power demand from the drive motor, or to simply reduce flow rate for a particular application. The effect of partially closing a discharge valve is called throttling and is illustrated in Figure 5.6. Initially, the operating point is at point A, corresponding to the intersection of the pump head curve and system head curve. The flow rate is Q_1 and the head developed by the pump is H_1. Suppose for some reason we need to reduce the flow rate to Q_2, which corresponds to point B on the pump head curve. By partially closing the discharge valve, we are in effect introducing additional pressure drop in the discharge piping such that the system head curve shifts to the left, as shown by the dashed curve that intersects the pump head curve at point B, corresponding to the capacity Q_2. At the flow rate of Q_2, the original system head curve shows a head requirement of H_3, corresponding to point C, whereas at a capacity of Q_2, the pump

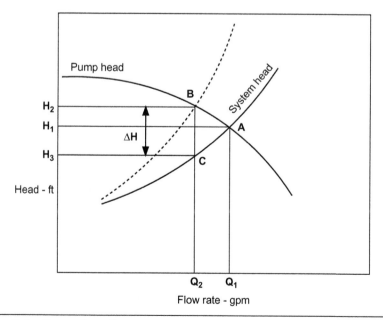

Figure 5.6   Pump throttling.

head available is H_2, corresponding to point B. Therefore, by using the discharge valve, we are throttling the pump head by $\Delta H = H_2\text{-}H_3$.

The throttled head ΔH, and thus the throttled pressure (in psi), represents a wasted pump head and correspondingly wasted energy. The pump throttling causes the system head curve to shift to an artificial head curve, shown by the dashed system head curve. The power lost in throttling can be calculated using Equation (2.5) as follows:

$$\text{Power lost in throttling} = Q_2 \times \Delta H \times Sg/(3960 \times E_2),$$

where E_2 is the pump efficiency at capacity Q_2, and Sg is the liquid specific gravity.

For example, if $Q_2 = 1800\,\text{gpm}$, $E_2 = 72\%$, and $\Delta H = 200\,\text{ft}$, $Sg = 0.85$:

$$\text{Power lost in throttling} = 1800 \times 200 \times 0.85/(3960 \times 0.72) = 107.32\,\text{HP}$$

If the pump were operated in the throttled mode continuously for a month, the cost of throttling at $0.10 per kWh is

$$\text{Wasted power cost} = 107.32 \times 0.746 \times 24 \times 30 \times 0.10 = \$5,764 \text{ per month}$$

It can be seen that throttling causes wasted energy and money and hence must be avoided as much as possible. Instead of running the pump in the throttled mode, consideration should be given to trimming the pump impeller to match the head required by the system at the lower flow rate.

EXAMPLE 5.1 USCS UNITS

A pump at a refinery is used to transport gasoline and diesel to a terminal 10 miles away, as shown in Figure 5.7. The pump is fed at a constant suction pressure of 50 psi. The origin location is at an elevation level of 100 ft. The terminus is at an elevation of 150 ft. The interconnecting pipeline is 16 inches with 0.250 inch wall thickness. Assuming a maximum operating pressure of 500 psi, calculate the maximum flow rates possible when pumping the two products separately. Select a suitable pump for this application. Once the pump is installed, determine the range of flow rates possible. Use a minimum pressure of 50 psi at the terminus. The absolute roughness of pipe is 0.002 inch. The properties of the two products are as follows:

Gasoline: $Sg = 0.74$ and viscosity $= 0.6\,\text{cSt}$
Diesel: $Sg = 0.85$ and viscosity $= 5.0\,\text{cSt}$

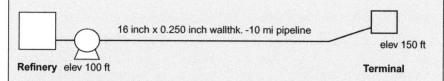

Figure 5.7   Pumping two products from refinery to terminal.

Solution

Since the maximum pressure is limited to 500 psi, we have to develop the system curves to determine the maximum flow rate for each product without exceeding the given pressure. Choose a range of flow rates from 1000 to 6000 gal/min. Using the Colebrook-White equation, we calculate for each product the pressure required at the pump location for each of these flow rates. The method will be explained in full for one flow rate, and the procedure will be repeated for the others.

Gasoline: $Q = 1000\,\text{gal/min}$

$$R = (3160 \times 1000)/(0.6 \times 15.5) = 339{,}785$$

$$1/\sqrt{f} = -2\,\text{Log}_{10}\left[(0.002/3.7 \times 15.5) + 2.51/(339785 \times \sqrt{f})\right]$$

Solving by trial and error, we get friction factor $= 0.0154$. Using Equation (4.12), the pressure loss is

$$Pm = 71.1475 \times 0.0154 \times (1000)^2 \times 0.74/(15.5)^5 = 0.9063\,\text{psi/mi}$$

The discharge pressure required at the pump for 1000 gal/min flow rate is

$$P_{1000} = 0.9063 \times 10 + ((150 - 100) \times 0.74/2.31) + 50 = 75\,\text{psi}$$

Similarly, discharge pressure for the remaining flow rates are

$$P_{2000} = 100.0\,\text{psi}$$
$$P_{4000} = 194.0\,\text{psi}$$
$$P_{6000} = 347.0\,\text{psi}$$
$$P_{8000} = 560.0\,\text{psi}$$

From the discharge pressure calculations, we get the following table for the system curves:

Gasoline–System Head Curve

Q (gal/min)	1000	2000	4000	6000	8000
H (psi)	75	100	194	347	560

Note that the system head curve values in the table are in psi, not ft of head.

Similarly, we calculate the discharge pressure for diesel as well and tabulate as follows:

Diesel–System Head Curve

Q (gal/min)	1000	2000	4000	6000	8000
H (psi)	84	121	254	461	740

The system head curves for diesel and gasoline are plotted in Figure 5.8. From Figure 5.8, since the maximum pipeline pressure is limited to 500 psi, the maximum flow rate for gasoline is 7491 gal/min and for diesel 6314 gal/min. To determine the operating point, we must superimpose a pump curve on the system head curves. Since the pump head curve is in ft of liquid head, we first convert the pressures to ft of head and replot the system curves, as shown in Figure 5.9. Note that we have deducted the 50 psi suction head from the system head curve pressures calculated earlier for the two liquids.

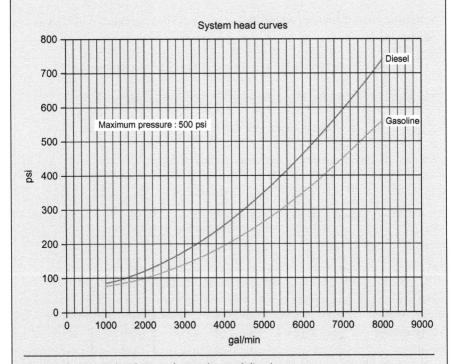

Figure 5.8   System head curves for gasoline and diesel.

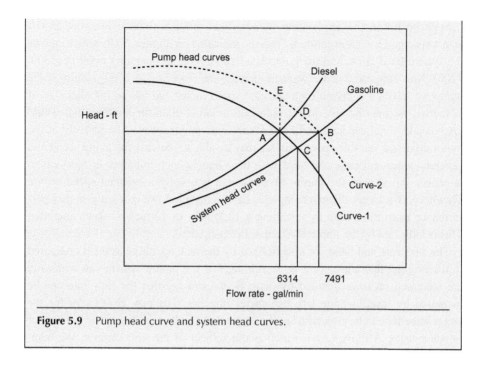

Figure 5.9   Pump head curve and system head curves.

Reviewing Figure 5.9, it is not possible to attain the maximum flow rate for each product with any selected pump. If we choose a suitable pump to achieve the maximum flow rate of 6314 gal/min with diesel, the operating point of the pump head, curve-1, will be at point A corresponding to this flow rate. Having selected this pump curve, the flow rate possible with gasoline is indicated by point C, the intersection of the pump head curve and gasoline system head curve. This flow rate is approximately 6900 gal/min, which is less than the desired maximum flow rate of 7491 gal/min. Alternatively, if we choose a pump head curve, shown as the dashed curve-2, suitable for pumping the maximum flow rate of 7491 gal/min of gasoline as indicated by the point B, the intersection of the gasoline system curve and the pump head curve-2. This pump head curve-2 intersects the diesel system curve at point D, which is at a higher pressure than the maximum 500 psi specified. Therefore, when pumping diesel, the flow rate will have to be cut back to 6314 gal/min to limit the pipeline pressure to 500 psi, using a discharge control valve. This would result in throttling the pump discharge pressure by an amount indicated by AE in the Figure 5.9. The use of pump head curve-1 to satisfy the required diesel flow rate of 6314 gal/min is a better option because no energy will be wasted in throttling as with pump head curve-2. Therefore, in conclusion, the pump selected in this case should satisfy the requirement of operating point A (6314 gal/min). The differential head required for the pump is (500 − 50) = 450 psi. Therefore, the pump differential head required is 450 × 2.31/0.85 = 1223 ft at a capacity of 6314 gal/min.

Suppose, for another application, we selected a pump to provide a head of 1200 ft at a capacity Q = 2300 gal/min. This is indicated in Figure 5.10, which shows the system head curve and the pump head curve with an operating point at (2300, 1200). Now suppose that we want to increase the flow rate to 2500 gal/min. If the pump we selected is driven by a constant speed motor and has an impeller size of 12 inches, we may be able to accommodate a larger-diameter impeller that would give us additional capacity and head at the same pump speed. Alternatively, if the motor drive is a variable speed unit, we may be able to increase the pump speed and generate additional flow and head with the same 12-inch impeller. In most cases, to reduce initial cost, the pump drive selected is usually a constant speed motor. Therefore, if a larger-diameter impeller can be installed, we can get a higher performance from this pump by installing a 12.5-inch- or 13-inch-diameter impeller. This is indicated by the upper H-Q curve in Figure 5.10.

The flow rate and head are now defined by the new operating point B compared to the original operating point A. Assuming that the piping system can withstand the increased pressure denoted by point B, we can say that the flow rate can be increased by installing the larger-diameter impeller. The new H-Q curve for the larger impeller can be created from the original H-Q curve for the 12-inch-diameter impeller using Affinity Laws, which is the subject of the next chapter. We mentioned earlier that an increase in capacity and head is possible by increasing the speed of the pump. This is also discussed under Affinity Laws in the next chapter.

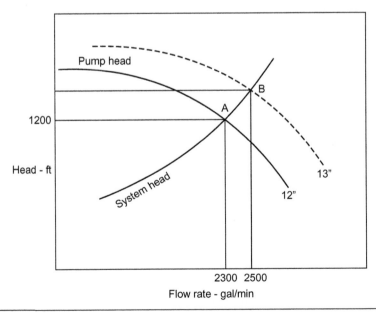

Figure 5.10 Increasing capacity with larger impeller.

For the present, we can state that once a pump is selected and installed to produce a certain head at a capacity Q, a further increase in flow rate may be achieved by increasing impeller diameter or increasing pump speed. Another, more expensive option would be to install a larger pump or add another pump in series or parallel arrangement. Multiple pump operation in series and parallel configuration is discussed in Chapter 8.

Types of System Head Curves

When we developed the system head curve in the previous example, we calculated the pressures required at various flow rates. At each flow rate we calculated the pressure drop due to friction, which was added to a constant component that represented the effect of the elevation difference between the pump location and pipeline terminus. The system head therefore consists of a frictional head component that depends on the flow rate and an elevation component that is constant. The shape of a system head curve may be flatter or steeper depending on which of the two components (friction versus elevation) is the larger value. Figure 5.11 shows two system curves as follows.

In Figure 5.11, the flatter system curve A has a smaller frictional head component and a larger elevation component. The system head curve B shows one that has a smaller elevation component but a larger friction head component. An example of system head curve A is one in which the pipeline elevation profile is fairly flat, but

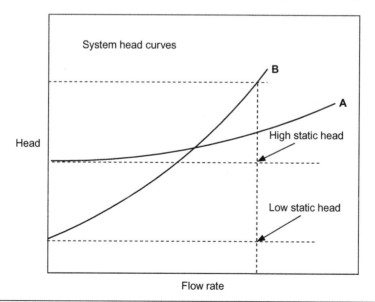

Figure 5.11   Types of system head curve.

the flow rates are high, contributing to higher frictional head loss. System curve type B occurs when the product is pumped to a high elevation point and the flow rates are fairly small. Example 5.2 illustrates the two different types of system head curves.

EXAMPLE 5.2 USCS UNITS

Consider two applications for a centrifugal pump depicted in Figure 5.12:

Application A: This is used for pumping water on an NPS 8 pipeline, 10 mi long. It requires pumping from a storage terminal at Compton (elevation: 100 ft) to a delivery terminus at Bailey (elevation: 150 ft) 10 miles away. The pipeline is 8.625 inches outside diameter, 0.250 inch wall thickness, pumping water at flow rates up to 3000 gal/min.

Application B: This is a 5-mile pipeline with a 16-inch pipe size. It requires pumping water from Beaumont (elevation: 250 ft) to a terminal at Denton (elevation: 780 ft) 5 miles away. The pipeline is 16-inch diameter and 0.250 inch wall thickness, with flow rates up to 4000 gal/min.

Use Hazen-Williams equation with C = 120. The terminus delivery pressure required is 50 psi. Develop a system head curve and compare the two applications.

Solution

Application A: Using Hazen-Williams equation (4.27) and rearranging to solve for head loss:

$$Q = 6.7547 \times 10^{-3} \times (C) \times (D)^{2.63} \, (h)^{0.54}$$

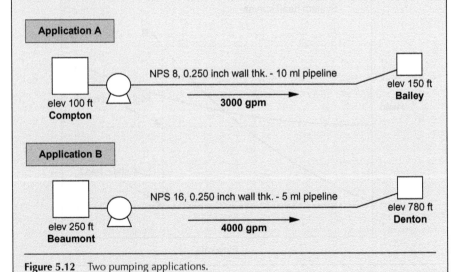

Figure 5.12   Two pumping applications.

Rearranging the equation to solve for h, we get

$$h = 10,357 \, (Q/C)^{1.85} \, (1/D^{4.87})$$

Considering flow rates ranging from 1000 to 4000 gal/min, we calculate the head loss for Application A as follows:

$$h_{1000} = 10,357(1000/120)^{1.85}(1/8.125)^{4.87} = 19.408 \text{ ft/1000 ft of pipe}$$

Therefore, total frictional head loss in 10 mi of pipe is

$$H_{1000} = 19.408 \times 10 \times 5.28 + \text{elevation head} + \text{delivery head}$$

$$= 1024.74 + (150 - 100) + (50 \times 2.31/1.0)$$

$$= 1024.74 + 50 + 115.5 = 1190 \text{ ft}$$

Similarly, we calculate the system head required for 2000 to 4000 gal/min as follows:

$$H_{2000} = 69.97 \times 52.8 + 50 + 115.5 = 3860 \text{ ft}$$

$$H_{3000} = 148.13 \times 52.8 + 50 + 115.5 = 7987 \text{ ft}$$

$$H_{4000} = 252.23 \times 52.8 + 50 + 115.5 = 13,483 \text{ ft}$$

Calculated values of head versus flow rates are tabulated as follows:

System Head Curve A

Q (gal/min)	1000	2000	3000	4000
H (ft)	1190	3860	7987	13,483

Considering flow rates ranging from 1000 to 4000 gal/min, we calculate the head loss for Application B as follows:

$$h_{1000} = 10,357 \, (1000/120)^{1.85} \, (1/15.5)^{4.87} = 0.8353 \text{ ft/1000 ft of pipe}$$

Therefore, total frictional head loss in 10 mi of pipe is

$$H_{1000} = 0.8353 \times 5 \times 5.28 + \text{elevation head} + \text{delivery head}$$

$$= 22.05 + (780 - 250) + (50 \times 2.31/1.0)$$

$$= 22.05 + 530 + 115.5 = 668 \text{ ft}$$

Similarly, we calculate the system head required for 2000 to 4000 gal/min as follows:

$$H_{2000} = 3.0113 \times 26.4 + 530 + 115.5 = 725 \, \text{ft}$$

$$H_{3000} = 6.3755 \times 26.4 + 530 + 115.5 = 814 \, \text{ft}$$

$$H_{4000} = 10.8556 \times 26.4 + 530 + 115.5 = 932 \, \text{ft}$$

Calculated values of head versus flow rates are tabulated as follows:

System Head Curve B

Q (gal/min)	1000	2000	3000	4000
H (ft)	668	725	814	932

System head curve A and B are plotted as shown in Figure 5.13.

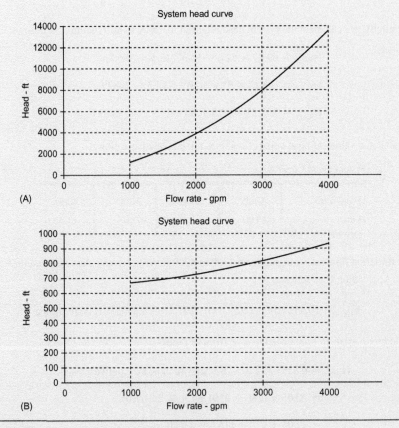

Figure 5.13 Comparison of system head curves.

EXAMPLE 5.3 SI UNITS

A centrifugal pump located at a storage facility at Anaheim is used to transport water from a large storage tank at an elevation of 150 m to a distribution terminal in the town of San Jose (elevation: 450 m), 25 km away, as shown in Figure 5.14. Calculate the maximum flow rate possible and select a suitable pump based on the following requirements: maximum pipe pressure: 5600 kPa; delivery pressure at San Jose: 300 kPa; and pipe is 500 mm outside diameter with a wall thickness of 12 mm. The Hazen-Williams C factor is 110. It is expected that the pump suction will be located at an elevation of 14 m below the bottom of the tank at Anaheim and at a distance of 12 m away. The suction piping from the tank is 600 mm outside diameter with a wall thickness of 10 mm.

There are two DN600 gate valves and two DN600 LR elbows in the suction piping. On the discharge side of the pump, there are two DN 500 gate valves and four DN500 LR elbows. In addition, the meter manifold on the discharge of the pump may be assumed to have a pressure drop of 100 kPa. At the delivery terminus at San Jose, an incoming meter manifold may be assumed to have a $\Delta P = 80$ kPa.

Solution

Since the maximum pressure is limited to 5600 kPa, we have to first determine the maximum permissible pressure drop in the piping from the pump location at Anaheim to the delivery point at San Jose 25 km away. The total pressure drop is the sum of the pressure drop in the 25 km length of DN500 pipe and the pressure drop through the two meter manifold and the minor losses through the valves and elbows.

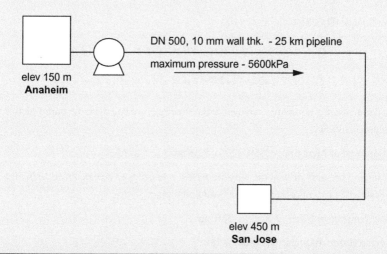

Figure 5.14   Anaheim to San Jose pipeline.

Using an L/D ratio of 8 for the gate valves and L/D = 16 for a LR elbow, we calculate the total equivalent length as follows:

Total equivalent length = $25 \, km + (2 \times 8 \times 0.5 + 4 \times 16 \times 0.5)/1000 \, km$

$$L_{eq} = 25.04 \, km$$

If the pressure drop is P_{km} kPa/km, the total pressure required is

$$P_{tot} = P_{km} \times 25.04 + 100 + 80 + 300 + (450 - 115) \times 1.0/0.102,$$

where the elevation difference between the pump location (115 m) and the delivery point (450 m) has been converted to pressure in kPa:

$$P_{tot} = (25.04 P_{km} + 480 + 3284.31) \, kPa$$

This total pressure at the discharge of the Anaheim pump must not exceed the maximum pressure of 5600 kPa. Therefore,

$$5600 = 25.04 \, P_{km} + 3764.31$$

Solving for the pressure drop permissible, we get

$$P_{km} = 73.31 \, kPa/km$$

From the pressure drop, we calculate the flow rate using the Hazen-Williams equation, as follows:

Inside diameter D = $500 - 2 \times 12 = 476 \, mm$

From Equation (4.28), using C = 110

$$Q = 9.0379 \times 10^{-8} \, (110)(476)^{2.63}(73.31/1.0)^{0.54} \, m^3/h$$

$$Q = 1113.7 \, m^3/h$$

In order to select a pump that can handle the flow rate, we calculate the pump suction pressure as follows:

Suction head at Anaheim = $(150 - 136) = 14 \, m$

The suction pressure drop at the maximum flow rate needs to be calculated, using the flow rate of $1113.7 \, m^3/h$ in the DN 600 suction piping:

Inside diameter = $600 - 2 \times 10 = 580 \, mm$

From the Hazen-Williams equation (4.28):

$$1,113.7 = 9.0379 \times 10^{-8} \, (110)(580)^{2.63}(P_{km}/1.0)^{0.54}$$

where P_{km} is the pressure loss in the DN 600 suction piping.

Solving for P_{km}, we get

$$P_{km} = 27.91 \text{ kPa/km}$$

Converting kPa to head, using Equation (1.12):

$$P_{km} = 27.91 \times 0.102/1.0 = 2.847 \text{ m/km}$$

The total equivalent length of the suction piping is next calculated taking into account the two gate valves and two LR elbows:

$$L = 12 \text{ m} + (2 \times 8 \times 0.6 + 2 \times 16 \times 0.6) \text{ m} = 40.8 \text{ m}$$

The total pressure drop in the suction piping is therefore

$$2.847 \times 40.8/1000 = 0.1162 \text{ m}$$

The suction head at the pump is therefore

$$14 \text{ m} - 0.1162 = 13.88 \text{ m}$$

Converting to pressure, the pump suction pressure is

$$Psuct = 13.88 \times 1.0/.102 = 136.08 \text{ kPa}$$

The pump differential pressure required is

$$\Delta P = Pdisch - Psuct = 5600 - 136.08 = 5464 \text{ kPa}$$

Converting to pump head:

$$H = 5464 \times 0.102/1 = 558 \text{ m, rounding up}$$

Thus, the pump required at Anaheim is

$$Q = 1113.7 \text{ m}^3/\text{h} \quad H = 558 \text{ m}$$

The power required to drive the pump at the design point can be estimated using a pump efficiency of 80% as follows:

From Equation (2.6): power $= 1113.7 \times 558 \times 1.0/(367.46 \times 0.8) = 2114 \text{ kW}$.
The motor power required at 95% motor efficiency $= 2114/0.95 = 2226 \text{ kW}$.
The electric motor driver must be at least $1.1 \times 2226 = 2448 \text{ kW}$.
The nearest standard size electric motor with a nameplate rating of 2500 kW would be
adequate for this application.

Summary

In this chapter the system head curve was introduced and the method of developing the system head curve for a pipeline was explained in detail. The operating point of a pump curve as the point of intersection between the H-Q curve for the pump and the system head curve for the pipeline was illustrated using an example. When two different products are pumped in a pipeline, the nature of the system head curves and their points of intersection with the pump head curve were reviewed. An example problem considering two products was used to illustrate the difference between two system curves. The steep system curve and the flat system curve and when these occur were discussed. Examples of system curve development in USCS units and SI units were shown. In the next chapter we will discuss the Affinity Laws for centrifugal pumps and how the performance of a pump at different impeller diameters and impeller speeds may be calculated.

Problems

5.1 A refinery pump is used to transport jet fuel and diesel to a terminal 20 km away, similar to Figure 5.6. The pump takes suction at 4.0 bar and is located at an elevation of 25 m. The tank at the terminus is at an elevation of 52 m. The pipeline between the refinery and the terminus consists of NPS 300, 6 mm wall thickness pipe, with a maximum operating pressure of 40 bar. Calculate the maximum flow rates possible when pumping the two products separately. Based on this, select a suitable pump. With the selected pump determine the range of flow rates possible with the two products. The delivery pressure at the terminus must be a minimum of 5 bar. Use the Hazen-Williams equation with C = 110 for jet fuel and C = 100 for diesel.

The properties of the two products are as follows:
   Jet: Sg = 0.804 and viscosity = 2.0 cP
   Diesel: Sg = 0.85 and viscosity = 5.0 cSt

5.2 A water storage tank at Denton is at an elevation of 150 ft. The water level is 20 ft. A pump located 20 ft away from a tank and at an elevation of 130 ft is used to pump water through a NPS 20 pipe, 0.500 in. wall thickness, to a distribution tank in the middle of the city of Hartford 55 mi away. The tank at Hartford is located at an elevation of 220 ft. The pipeline pressure is limited to 600 psig. The suction piping from the Denton tank to the pump consists of 20 ft of NPS 24 pipe, 0.500 in. wall thickness along with two gate valves and four 90-degree elbows, all NPS 24. On the discharge side of the pump there are two NPS 20 gate valves, one NPS 20 check valve, and six NPS 20 90-degree elbows. Calculate the maximum flow rate that can be achieved. Select a suitable pump for this application.

5.3 Crude oil (Sg = 0.89 and viscosity = 25 cSt at 20°C) is pumped from a valley storage tank (elevation 223 m above MSL) to a refinery on top of a hill (elevation 835 m above MSL), 23 km away using a DN 350, 8 mm wall thickness pipe. It is proposed to use two pumps to handle the task of pumping the crude oil at a flow rate of 1150 m³/h. Develop the system head curve for flow rates ranging from 500 to 1500 m³/h. Select suitable pumps for this application.

Pump Performance at Different Impeller Sizes and Speeds

In this chapter we will discuss the Affinity Laws for centrifugal pumps. We will use the affinity laws to predict the performance of a pump at different impeller diameters or speeds. Also, the method of calculating the speed or impeller trim necessary to achieve a specific pump operating point will be reviewed and explained using examples.

The Affinity Laws for centrifugal pumps are used to determine the performance of a pump at different impeller diameters or speeds. For example, if we are given the H-Q curve for a pump with an impeller diameter of 12 inches running at 3560 RPM, we can predict the new performance at a larger or smaller impeller diameter or at a lower or higher pump speed. This is illustrated in Figure 6.1, where the head-capacity

DOI: 10.1016/B978-1-85617-828-0.00006-8

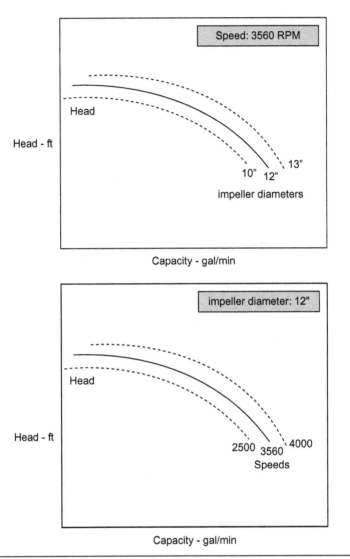

Figure 6.1 Pump performance at different size or speed.

curves for a particular pump are shown at various impeller diameters and at various speeds. In each case, either the pump speed or the impeller diameter is kept constant, while varying the other.

It can be seen from Figure 6.1 that within a pump casing a range of impeller sizes can be installed, and the pump vendor can provide a family of pump head curves corresponding to these impeller diameters. Similarly, for a fixed impeller diameter, there is a family of head curves that correspond to the range of permissible pump speeds.

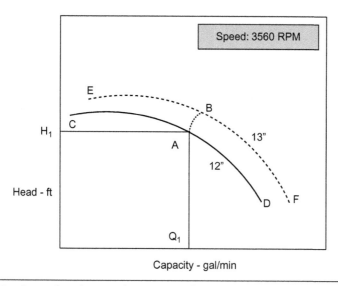

Speed: 3560 RPM

Figure 6.2   Pump head curve for increased impeller diameter.

Consider a pump fitted with a 12-inch impeller running at 3560 RPM, as indicated in Figure 6.2. This head curve CD has a range of capacities between points C and D. At the point A, the capacity is Q_1, and the head available is H_1.

Suppose that instead of the 12-inch impeller, a larger impeller 13 inches in diameter is installed in this pump. The H-Q curve for the larger impeller is the upper dashed curve EF, indicating that both the head and the capacity are increased by some factor and the curve shifts to the top and to the right, compared to that of the 12-inch impeller. Points E and F both represent slightly higher capacities compared to those at C and D on the original 12-inch impeller head curve.

For each Q value on the 12-inch curve there is a corresponding Q value on the 13-inch curve defined by the following Affinity Law for centrifugal pumps:

$$Q_2/Q_1 = D_2/D_1 \qquad (6.1)$$

where Q_1 corresponds to point A (impeller diameter D_1), and Q_2 corresponds to point B (impeller diameter D_2).

Setting $D_1 = 12$ and $D_2 = 13$, the capacity ratio becomes

$$Q_2/Q_1 = 13/12 = 1.0833$$

Thus, every point on the lower 12-inch diameter curve with a certain Q value has a corresponding Q value on the 13-inch diameter curve that is 8.33% higher.

The head, on the other hand, bears the following relationship with diameter change:

$$H_2/H_1 = (D_2/D_1)^2 \qquad (6.2)$$

Equations (6.1) and (6.2) are called the Affinity Laws equation for pump impeller diameter change. The impeller diameter change assumes the pump speed is kept constant. Instead of diameter change, if the speed of the impeller were increased from $N_1 = 3560$ RPM to $N_2 = 4000$ RPM, the capacity Q and head H on the new H-Q curve are related by the Affinity Laws for speed change as follows:

$$Q_2/Q_1 = N_2/N_1 \qquad (6.3)$$

$$H_2/H_1 = (N_2/N_1)^2 \qquad (6.4)$$

In the present case, $N_2/N_1 = 4000/3560 = 1.1236$.

Thus, every Q value on the 3560 RPM curve is increased by 12.36% on the 4000 RPM curve as shown in Figure 6.3. The head values are increased by the factor $(1.1236)^2 = 1.2625$, and thus the heads are increased by 26.25%.

From Equation (2.3), the power required by the pump is proportional to the product of capacity Q and the head H. Therefore, we have the following Affinity Laws equation for the change in power required when impeller diameter or speed is changed:

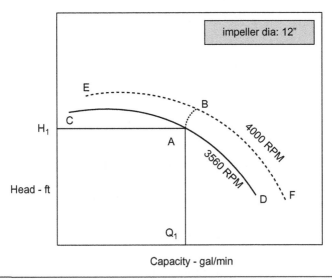

Figure 6.3 Pump head curve for increased speed.

For diameter change:

$$BHP_2/BHP_1 = (D_2/D_1)^3 \tag{6.5}$$

For speed change:

$$BHP_2/BHP_1 = (N_2/N_1)^3 \tag{6.6}$$

Thus, the pump capacity Q varies directly with the impeller diameter or speed ratio. The pump head H varies as the square of the ratio, and the pump power varies as the cube of the ratio, as expressed by the Affinity Laws equations (6.1) through (6.6).

It is seen that using Affinity Laws, we can quickly predict the performance of a larger or smaller impeller given the performance at any other impeller diameter. The minimum and maximum impeller sizes possible in a pump depend on the pump casing size and are specified by the pump manufacturer. Therefore, when calculating the performance of trimmed impellers, using Affinity Laws, the pump vendor must be consulted to ensure that we are not considering impeller sizes beyond practical limits for the pump.

The Affinity Laws are considered to be only *approximately* true for diameter changes. However, they are exactly correct when applied to speed changes. Furthermore, when predicting the performance of a trimmed impeller using Affinity Laws, we may have to apply a correction factor to the H and Q values for the calculated diameter. Consult the pump manufacturer to obtain these correction factors. The approximate correction factors for diameter changes may be obtained from Figure 6.4.

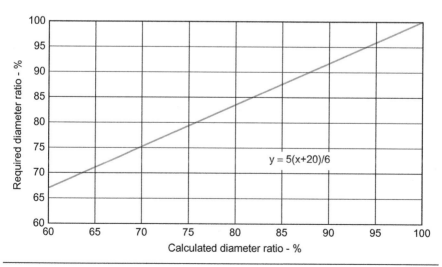

Figure 6.4 Correction factor for trimming impeller.

Note that the correction for impeller trim is a straight line, with the following equation:

$$y = (5/6)(x + 20) \tag{6.7}$$

where

x = calculated trim for impeller diameter, %

y = corrected trim for impeller diameter, %

For example, if the calculated trim from the affinity laws is 80%, we must correct this to the following value, using Figure 6.4.

Corrected impeller trim $y = (5/6)(80 + 20) = 83.33\%$

We have discussed how the H and Q values change with impeller diameter and speed. The efficiency curve versus capacity can be assumed to be approximately same with impeller diameter change and speed change. An example will illustrate the use of the Affinity Laws.

EXAMPLE 6.1 USCS UNITS

The following centrifugal pump curve is for a two-stage, 12-inch impeller at a speed of 3560 RPM:

PUMP61 performance for diameter = 12 inch speed = 3560 RPM

Q (gal/min)	0	800	1200	1600	2000	2400
H (ft)	1570	1560	1520	1410	1230	1030
E (%)	0	57.5	72.0	79.0	79.8	76.0

a. Determine the performance of this pump at 13-inch impeller diameter, keeping the speed constant at 3560 RPM.

b. Keeping the diameter at 12 inches, if the speed is increased to 4000 RPM, determine the new H-Q curve at the higher speed.

Solution

a. Using Affinity Laws equations (6.1) and (6.2) for diameter change, keeping the speed constant, we create a new set of Q-H data by multiplying the Q values given by the factor 13/12 = 1.0833 and the H values by a factor of $(1.0833)^2 = 1.1735$ as follows:

At $Q_1 = 800$ $Q_2 = 800 \times 1.0833 = 866.64$

At $H_1 = 1560$ $H_2 = 1560 \times (1.0833)^2 = 1830.72$

Other values of Q and H are similarly calculated, and the following table of Q, H values is prepared for the 13-inch diameter impeller.

PUMP61 performance for diameter = 13-inch speed = 3560 RPM

Q (gal/min)	0	866.64	1299.96	1733.28	2166.6	2599.92
H (ft)	1842.46	1830.72	1783.78	1654.69	1443.45	1208.75
E (%)	0	57.5	72.0	79.0	79.8	76.0

Note that the efficiency values remain the same as for the original 12-inch impeller. The H-Q curves for the 12-inch and 13-inch diameters are plotted as shown in Figure 6.5.

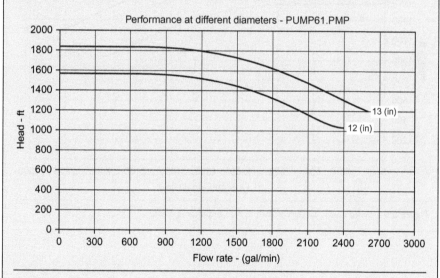

Figure 6.5   Head capacity curves for the two impeller diameters.

b.   When the speed is increased from 3560 to 4000 RPM, with 12-inch diameter kept constant, using the Affinity Laws equations (6.3) and (6.4), we get the new values of Q and H as follows:

At $Q_1 = 800$ $Q_2 = 800 \times (4000/3560) = 800 \times 1.1236 = 898.88$

$H_1 = 1560$ $H_2 = 1560 \times (4000/3560)^2 = 1969.46$

Other values of Q and H are similarly calculated, and the following table of Q, H values is prepared for the 12-inch diameter impeller:

PUMP61 performance for diameter = 12-inch speed = 4000 RPM

Q (gal/min)	0.0	898.88	1348.32	1797.76	2247.2	2696.64
H (ft)	1982.09	1969.46	1918.97	1780.09	1552.85	1300.35
E (%)	0.0	57.5	72.0	79.0	79.8	76.0

Note that the efficiency values remain the same as for the original 3560 RPM speed. The H-Q curves for 3560 RPM and 4000 RPM are shown in Figure 6.6.

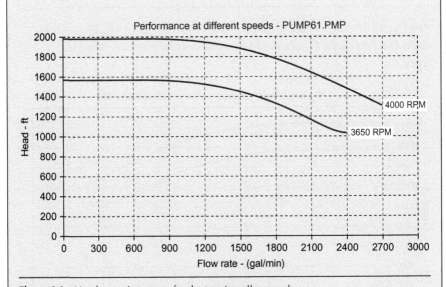

Figure 6.6 Head capacity curves for the two impeller speeds.

EXAMPLE 6.2 SI UNITS

The following pump curve data are for a 250 mm diameter impeller at a speed of 2950 RPM.

PUMP62 performance for diameter = 250 mm speed = 2950 RPM

Q (m³/h)	0	1000	1500	2000	2500
H (m)	300	275	245	210	165
E (%)	0	65	70	72	65

a. Determine the performance of this pump at 230 mm impeller diameter, keeping the speed constant at 2950 RPM.

b. Keeping the diameter at 250 mm, if the speed is increased to 3200 RPM, determine the new H-Q curve at the higher speed.

Solution

a. Using Affinity Laws equations (6.1) and (6.2) for diameter change, keeping the speed constant, we create a new set of Q-H data by multiplying the Q values given by the factor $230/250 = 0.92$ and the H values by a factor of $(0.92)^2 = 0.8464$ as follows:

$$\text{At } Q_1 = 1000 \qquad Q_2 = 1000 \times 0.92 = 920.0$$
$$H_1 = 275 \qquad H_2 = 275 \times (0.92)^2 = 232.76$$

Other values of Q and H are similarly calculated, and the following table of Q, H values is prepared for the 230 mm diameter impeller:

PUMP62 performance for diameter = 230 mm speed = 2950 RPM

Q (m³/h)	0	920	1380	1840	2300
H (m)	253.92	232.76	207.37	177.74	139.66
E (%)	0	65	70	72	65

Note that the efficiency values remain the same as for the original 250 mm impeller. The H-Q curves for 250 mm and 230 mm diameters are plotted as shown in Figure 6.7.

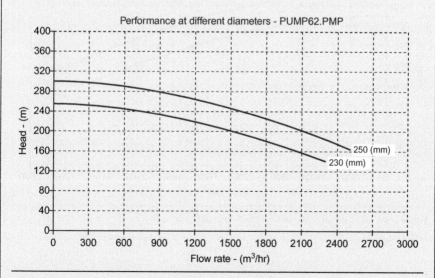

Figure 6.7 Head capacity curves for the two impeller diameters.

b. When the speed is increased from 2950 to 3200 RPM, with 250 mm diameter kept constant, using the Affinity Laws equations (6.3) and (6.4), we get the new values of Q and H as follows:

At $Q_1 = 1000$ $Q_2 = 1000 \times (3200/2950) = 1000 \times 1.0847 = 1084.7$
At $H_1 = 275$ $H_2 = 275 \times (3200/2950)^2 = 323.59$

Other values of Q and H are similarly calculated, and the following table of Q, H values is prepared for the 250 mm diameter impeller:

PUMP62 performance for diameter = 250 mm speed = 3200 RPM

Q (m³/h)	0	1084.7	1627.12	2169.49	2711.86
H (m)	353.0	323.59	288.29	247.10	194.15
E (%)	0	65	70	72	65

Note that the efficiency values remain the same as for the original 2950 RPM speed. The H-Q curves for 2950 RPM and 3200 RPM are shown in Figure 6.8.

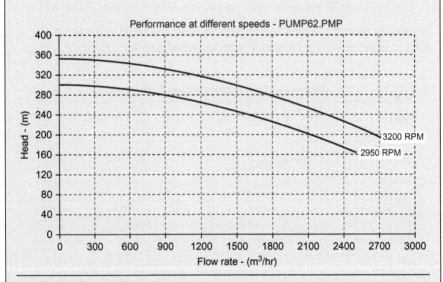

Figure 6.8 Head capacity curves for the two impeller speeds

Calculating the Impeller Diameter or Speed for a Specific Operating Point

Since we can predict the performance of a pump at different impeller speeds or diameters, we can also calculate the required diameter or speed to achieve a specific operating point on the pump H-Q curve. Consider a pump H-Q curve (curve-1) shown in Figure 6.9. This pump has an impeller diameter of 12 inches and runs at 3560 RPM. The BEP is at point A, corresponding to the following data:

$$Q = 2000 \quad H = 1800 \text{ ft} \quad \text{and} \quad E = 82\%$$

Suppose for a particular application, we need to determine if this pump is suitable for the following operating condition:

$$Q = 1900 \quad H = 1680 \text{ ft}$$

It can be seen that the new operating point is at lower Q and H values. At the lower capacity Q = 1900, the head generated will be higher than that at the BEP. Therefore, we will try to obtain a new pump curve (curve-2) that will lie below the present curve shown as a dashed curve. This H-Q curve is one that passes through the point Q = 1900 and H = 1680. The dashed curve can be obtained by either reducing the impeller diameter or the speed.

Suppose we decide to trim the impeller diameter to obtain the new H-Q curve that satisfies the operating condition of Q = 1900 gal/min, H = 1680 ft. We could

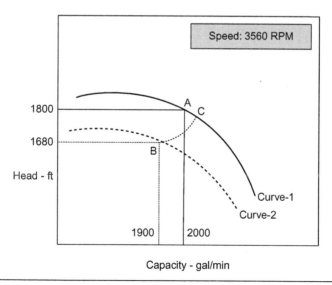

Figure 6.9 Requirement for new operating point.

progressively try smaller impeller sizes and create the new Q-H values using the Affinity Laws and finally arrive at the correct trimmed impeller size for the new curve. But this is quite a laborious process if it is done manually. A computer program or an Excel spreadsheet approach would help immensely in the process.

On an Excel spreadsheet the following approach will give the required impeller trim. The given pump curve data Q-H values are tabulated and an initial ratio of diameter of 0.9 is selected. Using this ratio, a new table of Q-H values is created using the Affinity Laws. By inspection, the 0.9 ratio will have to be increased to 0.96 as the next approximation. The process is repeated until the desired operating point (Q = 1900, H = 1680) is achieved.

Another analytical approach to determining the impeller trim is as follows. The given pump curve data is fitted to a polynomial, as described in Chapter 2, Equation (2.1):

$$H = a_0 + a_1 Q + a_2 Q^2$$

where a_0, a_1, and a_2 are constants for the pump. First, the given pump curve data (BEP: Q = 2000, H = 1800) is fitted to an equation, and the coefficients a_0, a_1, and a_2 are determined by using the least squares method.

There is a point C (Q_1, H_1) on the original head curve-1 that corresponds to the desired operating point B (Q_2 = 1900, H_2 = 1680) on the trimmed pump curve-2, as shown in Figure 6.9. These two points are related by the Affinity Laws for impeller diameter change, as follows:

$$Q_1/Q_2 = D_1/D_2 = r$$

$$H_1/H_2 = (D_1/D_2)^2 = r^2$$

where D_1 is the impeller diameter for curve-1 and D_2 is the impeller diameter for curve-2.

Solving for Q_1 and H_1 in terms of the given values of Q_2 and H_2, we get

$$Q_1 = 1900r \text{ and } H_1 = 1680\, r^2 \tag{6.8}$$

Note that r is the ratio D_1/D_2, which is a number greater than 1.0, since D_2 is the trimmed diameter, corresponding to curve-2, hence $D_2 < D_1$.

Since the point (Q_1, H_1) lies on the head curve-1, we can write

$$H_1 = a_0 + a_1 Q_1 + a_2 Q_1^2 \tag{6.9}$$

Substituting the values of Q_1 and H_1 from Equations (6.8) and (6.9), we get the following quadratic equation in r:

$$1680\ r^2 = a_0 + a_1(1900r) + a_2(1900r)^2 \tag{6.10}$$

Since a_0, a_1, and a_2 are known constants, we can solve for the diameter ratio r from Equation (6.7). The next example illustrates this method.

EXAMPLE 6.3 USCS UNITS

The following pump curve data is given for a 12-inch impeller running at 3560 RPM:

Q (gal/min)	0	1000	2000	2500	3000
H (ft)	2250	2138	1800	1547	1238

An application requires the following operating point Q = 1900, H = 1680. Determine how much the current impeller should be trimmed to achieve the desired operating condition.

Solution

First, we will fit a second-degree polynomial to the given H-Q data using the least squares method. The H-Q curve will have the following equation:

$$H_1 = a_0 + a_1 Q_1 + a_2 Q_1^2 \tag{6.11}$$

where a_0, a_1, and a_2 are constants for the pump curve of a specific impeller diameter and speed.

The coefficients are determined using the least squares method as follows:

$$a_0 = 2250.1 \qquad a_1 = 4.0512 \times 10^{-5} \qquad a_2 = -1.1249 \times 10^{-4}$$

Substituting these values in Equation (6.10), we get

$$1680\ r^2 = 2250.1 + 4.0512 \times 10^{-5}(1900r) - 1.1249 \times 10^{-4}(1900r)^2$$

Solving this quadratic equation in r, we get the diameter ratio D_1/D_2 as r = 1.0386. Therefore, 1/r = 1/1.0386 = 0.9628, or the trimmed impeller is 96.28% of the original diameter. This is the theoretical impeller trim required. Applying the correction factor for impeller trim, using Equation (6.7):

$$y = (5/6)(96.28 + 20) = 96.9\%$$

Therefore, the desired operating point (Q = 1900, H = 1680) can be achieved by trimming the current 12-inch impeller to 12.0 × 0.969 = 11.63 inches.

EXAMPLE 6.4 SI UNITS

The following pump curve is for a 250 mm diameter impeller at a speed of 1780 RPM:

Q (m³/h)	180	360	420	480
H (m)	380	310	278	235
E (%)	62	82	81	77

This pump is fitted with a variable speed drive with a speed range of 1500 RPM to 3000 RPM, and the following operating point is desired: Q = 450 m³/h, H = 300 m. Determine the speed at which the required condition can be achieved.

Solution

It is clear that the desired operating point can only be achieved by increasing the speed of this pump above the 1780 RPM, since at this original speed, the head generated at Q = 450 m³/h is less than 300 m. As in the previous example, we will fit a second-degree polynomial to the given H-Q data using the least squares method. The H-Q curve will have the following form from Equation (6.11):

$$H = a_0 + a_1 Q_1 + a_2 Q_1^2 \qquad (6.12)$$

where a_0, a_1, and a_2 are constants for the pump curve of a specific impeller diameter and speed.

The coefficients are determined using the least squares method as follows:

$$a_0 = 395.42 \qquad a_1 = 6.1648 \times 10^{-2} \qquad a_2 = -8.2182 \times 10^{-4}$$

There is a point (Q_1, H_1) on the original head curve-1 that corresponds to the desired operating point $(Q_2 = 450, H_2 = 300)$ on the increased speed pump curve-2. These two points are related by the Affinity Laws for impeller speed change, as follows:

$$Q_1/Q_2 = N_1/N_2 = r$$

$$H_1/H_2 = (N_1/N_2)^2 = r^2$$

Note that the speed ratio r is a number less than 1, since $N_1 < N_2$. Therefore, $Q_1 = 450r$ and $H_1 = 300r^2$.

Substituting these values in Equation (6.12), we get

$$300\, r^2 = 395.42 + 6.1648 \times 10^{-2}(450r) - 8.2182 \times 10^{-4}(450r)^2$$

Solving this quadratic equation in r, we get the speed ratio N_1/N_2 as r = 0.9510. Therefore, $1/r$ = 1/0.951 = 1.0516, or the increased impeller speed is 105.16% of the original speed of 1780 RPM.

Therefore, the desired operating point (Q = 450, H = 300) can be achieved by increasing the pump impeller speed to 1780 × 1.0516 = 1872 RPM.

Summary

In this chapter we introduced the Affinity Laws for centrifugal pumps. The method of determining the pump performance with changes in impeller diameter or impeller speed was illustrated with examples. Given a pump curve, the necessary impeller trim or change in speed required to achieve a specific operating point was explained using examples. In the next chapter, NPSH and pump cavitation and calculation of the NPSH available in a piping system will be analyzed in detail.

Problems

6.1 A pump with an impeller size of 10 inches running at 3570 RPM has the following characteristics:

Q gpm	0	2136	4272	5340	6408
H ft	4065	3862	3252	2795	2236
E%	0.0	63.8	85.0	79.7	63.8

a. Determine the new H-Q curve for this pump if the speed is decreased to 2800 RPM.

b. If the speed is kept constant at 3570 RPM but the impeller is trimmed to 9.5 inches, what is the new H-Q curve?

6.2 A pump with an impeller size of 300 mm running at 3560 RPM has the following characteristics:

Q L/s	0	18	36	45	54
H m	102	97	81	70	60
E%	0.0	63.8	85.0	79.7	63.8

It is desired to modify this pump at a constant speed such that the following operating condition is achieved: Q = 20 L/s H = 80 m

a. What should the impeller size be?

b. Instead of changing the impeller size, what speed should the pump be run at to achieve the same design point?

6.3 A pump with an impeller size of 9 inches running at 3560 RPM has the following characteristics:

Q gpm	0	992	1985	2481	2977
H ft	1166	1108	933	802	642
E%	0.0	62.0	84.0	79.0	64.0

The maximum and minimum impeller sizes possible within the pump casing are 11 inches and 8 inches, respectively. Determine the maximum capacity of this pump and the impeller sizes required to achieve the following conditions:

$$Q = 1000 \text{ gpm} \qquad H = 950 \text{ ft}$$

$$Q = 2100 \text{ gpm} \qquad H = 850 \text{ ft}$$

Chapter 7

NPSH and Pump Cavitation

The *net positive suction head* (NPSH) term was introduced in Chapter 2, where we discussed its variation with pump capacity. In this chapter we will discuss NPSH in more detail and explain its importance and its impact on pump cavitation. The terms *NPSH Required* ($NPSH_R$) and the *NPSH Available* ($NPSH_A$) will be reviewed, and the method of calculation of NPSH available in a pipeline configuration will be explained using examples.

NPSH represents the effective pressure at the centerline of the suction of a pump minus the vapor pressure of the liquid at the pumping temperature. It is a measure of how much the liquid pressure at the pump suction is above its vapor pressure. The smaller this number, the higher the danger of the liquid vaporizing and causing

DOI: 10.1016/B978-1-85617-828-0.00007-X

damage to the pump internals. Since a pump is designed to handle liquid only, the presence of vapors tends to cause implosions within the pump casing, resulting in damage to the impeller. Therefore, we must always ensure that the liquid pressure at the suction of the pump never falls below the vapor pressure of the liquid at the pumping temperature.

The minimum required NPSH for a particular pump is specified by the pump manufacturer. It is also referred to as NPSH Required ($NPSH_R$), and it increases with the increase in pump capacity, as shown in Figure 7.1. It represents the minimum required pressure at the pump suction for a particular pump capacity.

While the $NPSH_R$ for a particular pump depends on the pump design and is provided by the pump vendor, the available NPSH, designated as $NPSH_A$, depends on the way the pump is connected to the liquid storage facility. $NPSH_A$ is defined as the effective suction pressure at the centerline of the pump suction, taking into account the atmospheric pressure, tank head, vapor pressure of liquid at the flowing temperature, and the frictional head loss in the suction piping.

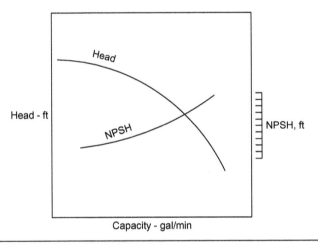

Figure 7.1 NPSH versus pump capacity.

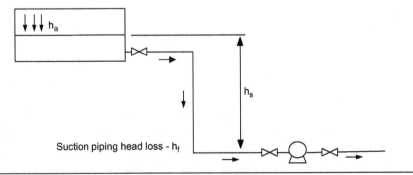

Figure 7.2 NPSH available.

Referring to Figure 7.2, the $NPSH_A$ can be calculated using the following equation:

$$NPSH_A = h_a + h_s - h_f - h_{vp} \tag{7.1}$$

where

$NPSH_A$: available net positive suction head, ft of liquid
h_a: atmospheric pressure at surface of liquid in tank, ft
h_s: suction head from liquid level in tank to pump suction, ft
h_f: frictional head loss in suction piping, ft
h_{vp}: vapor pressure of liquid at pumping temperature, ft

Sometimes the liquid is stored in a tank or other vessel under a pressure P that is higher than the atmospheric pressure. In such a case, the first term h_a in Equation (7.1) will be replaced with the absolute pressure P at the liquid surface, converted to ft of head of liquid. Note that NPSH is always expressed in absolute terms of pressure in ft of liquid in USCS units and in meters in SI units. We will illustrate the calculation of the available NPSH using an example.

EXAMPLE 7.1 SI UNITS

Referring to Figure 7.2, the water tank bottom is at an elevation of 10 m above mean sea level (MSL), and the water level in the tank is 5 m. The pump suction is located 2 m above the tank bottom. The following pipe, valves, and fittings comprise the suction piping from the tank to pump suction. Two DN 500 gate valves, four DN 500 90-degree LR elbows, and 22 m of DN 500 pipe, 10 mm wall thickness. For a flow rate of 1200 m³/h, calculate the available NPSH at the pump suction. Use the Hazen-Williams equation with C = 120. The vapor pressure of water at pumping temperature may be assumed to be 5 kPa absolute. Atmospheric pressure = 1 bar.

Solution

Calculate the suction piping head loss using Equation (4.28):

$$1200 = 9.0379 \times 10^{-8} \, (120)(500 - 2 \times 10)^{2.63} (P_{km}/1.0)^{0.54}$$

$$P_{km} = 68.81 \text{ kPa/km}$$

Calculate total equivalent lengths of fittings, valves, and pipe as follows:

2 – DN 500 gate valve = $2 \times 8 \times 480/1000 = 7.68$ m of DN 500 pipe
4 – DN 500 90° LR elbows = $4 \times 16 \times 480/1000 = 30.72$ m of DN 500 pipe
1 – 22 m of DN 500 pipe = 22 m of DN 500 pipe

Total length L = 7.68 + 30.72 + 22 = 60.4 m.

Therefore, the total head loss in the suction piping is

$$h_f = 68.81 \times 60.4/1000 = 4.16 \, \text{kPa}$$

From Equation (1.12), converting to head in m

$$= 4.16 \times 0.102/1.0 = 0.424 \, \text{m}$$

Calculating the terms in Equation (7.1) in $NPSH_A$:

$$h_a = 1 \times 100 \times 0.102 = 10.2 \, \text{m}$$

$$h_s = 5 - 2 = 3 \, \text{m}$$

$$h_f = 0.424 \, \text{m}$$

$$h_{vp} = 5 \times 0.102 = 0.51 \, \text{m}$$

The available NPSH using Equation (7.1) is

$$NPSH_A = h_a + h_s - h_f - h_{vp} = 10.2 + 3 - 0.424 - 0.51 = 12.27 \, \text{m}$$

Therefore, the pump used for this application should have an NPSH requirement of less than 12.27 m, at 1200 m³/h flow rate, to prevent cavitation of the pump.

EXAMPLE 7.2 USCS UNITS

Gasoline is stored in a tank at a pump station, and the tank is piped to a centrifugal pump located 50 ft away from the tank bottom, as shown in Figure 7.3. The liquid level in the tank is 20 ft. The tank bottom and pump suction are at elevations of 32 ft and 29 ft, respectively.

The suction piping consists of the following pipe, valves, and fittings: two NPS 16 gate valves, four NPS 16 90-degree LR elbows, and 50 ft of NPS 16 pipe, 0.250 inch wall thickness. Calculate the available NPSH at a flow rate of 2800 gpm. If the pump curve shows NPSH_R of 22 ft, will the pump cavitate? What needs to be done to prevent cavitation? Use the Colebrook-White equation with pipe absolute roughness = 0.002 inch and vapor pressure of gasoline at pumping temperature = 5.0 psia.

Sg of gasoline = 0.74, visc = 0.6 cSt, and atmospheric pressure = 14.7 psi

Solution
First, calculate the head loss in the NPS 16 suction piping using the Colebrook-White equation. Using Equation (4.8),

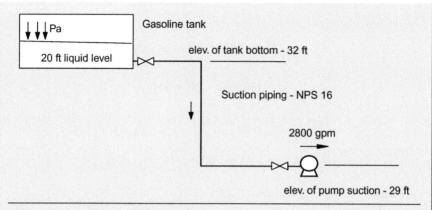

Figure 7.3 Calculation of NPSH available.

Reynolds number = 3160 × 2800/(0.6 × 15.5) = 951,398

Relative roughness = e/D = 0.002/15.5 = 0.000129

Calculate the friction factor f using the Colebrook-White equation (4.21):

$$1/\sqrt{f} = -2\,\text{Log}_{10}[(0.000129/3.7) + 2.51/(951,398\,\sqrt{f})]$$

Solving for f by trial and error, we get

$$f = 0.0139$$

The head loss from Equation (4.12) is

$$P_m = 71.1475 \times 0.0139\,(2800)^2 \times 0.74/(15.5)^5 = 6.41\,\text{psi/mi}$$

Calculate the equivalent lengths of fittings, valves, and pipe:

2 – NPS16 gate valves = 2 × 8 × 15.5/12 = 20.67 ft of NPS16 pipe

4 – NPS16 90° LR elbows = 4 × 16 × 15.5/12 = 82.67 ft of NPS16 pipe

1 – 50 ft of NPS16 pipe = 50 ft of NPS16 pipe

The total equivalent length of pipe, fittings and valves is

$$L = 20.67 + 82.67 + 50 = 153.34\text{ ft of NPS16 pipe}$$

Therefore, total head loss in suction piping is

$$h_f = 6.41 \times 153.34/5280 = 0.186\text{ psi}$$

$$= 0.186 \times 2.31/0.74 = 0.58\text{ ft}$$

Calculating the terms in Equation (7.1) in $NPSH_A$:

$$h_a = 14.7 \times (2.31/0.74) = 45.89 \, ft$$

$$h_s = (20 + 32 - 29) = 23 \, ft$$

$$h_f = 0.58 \, ft$$

$$h_{vp} = 5 \times (2.31/0.74) = 15.61 \, ft$$

The available NPSH, using Equation (7.1) is

$$NPSH_A = h_a + h_s - h_f - h_{vp} = 45.89 + 23 - 0.58 - 15.61 = 52.70 \, ft$$

Since $NPSH_R = 22 \, ft$ and is less than $NPSH_A$, the pump will not cavitate.

EXAMPLE 7.3 USCS UNITS

Three centrifugal pumps are installed in a parallel configuration at a pumping station to transport water from an atmospheric storage tank via a 35-inch inside diameter pipeline to a storage tank that is located 15 miles away. Each pump operates at a capacity of 6000 gpm and requires a minimum NPSH of 23 ft at the pumping temperature. The available NPSH has been calculated to be 30 ft. It is proposed to increase the capacity of the pipeline by 20%, by installing larger impellers in each pump. If the increased flow rate requires $NPSH_R$ to increase to 35 ft, but the available NPSH drops from 30 ft to 28 ft, what must be done to prevent cavitation of these pumps at the increased capacity?

Solution

Initially, at 6000 gpm flow rate through each pump, the available NPSH is 30 ft, compared to the $NPSH_R$ of 23 ft. Hence the pumps will not cavitate. When the flow rate in each pump increases by 20%, the $NPSH_R$ also increases to 35 ft. However, due to increased head loss in the suction piping at the higher flow rate, $NPSH_A$ decreases from 30 ft to 28 ft. Since this is less than $NPSH_R = 35 \, ft$, the pumps will cavitate at the higher flow rate.

If the flow rate cannot be reduced, we must provide additional positive pressure on the suction side of the pumps to ensure $NPSH_A$ is more than 35 ft. This is done by selecting a high-capacity, low-head (such as 80 to 100 ft) pump with a low NPSH requirement. A vertical can-type pump will be capable of handling this requirement. Sometimes it may be more economical to install multiple vertical tank booster pumps instead of a single large-capacity pump. Thus, an alternative would be to install three vertical can-type booster pumps in parallel, located close to the tank, each capable of providing a head of 80 to 100 ft at the increased flow rate. Assuming an efficiency of 75%, each of these booster pumps will have the following power requirement to handle 100 ft of head at the increased flow rate of $6000 \times 1.2 = 7200$ gpm:

$$BHP = (6000 \times 1.2) \times 100 \times 1.00/(3960 \times 0.75) = 243 \, HP$$

Considering 95% motor efficiency, a 300-HP drive motor will be required for each of these pumps. In conclusion, the NPSH problem can be solved by installing three parallel vertical booster pumps, each 300 HP, that can provide the 7200 gpm capacity at a head of 100 ft. With the booster pumps installed, the available NPSH is

$$NPSH_A = 100 + 28 = 128 \, ft,$$

which is more than the required NPSH of 35 ft. Hence, the pumps will not cavitate.

As mentioned before, insufficient NPSH will cause cavitation in pumps. Cavitation causes vaporization of the liquid in the pump casing or suction line. If the net pressure in the pump suction is less than the liquid vapor pressure, vapor pockets are formed. These vapor pockets reach the impeller surface and collapse, causing noise, vibration, and surface damage to the impeller. A damaged impeller loses efficiency and is expensive to operate, and it may cause further structural damage to the pump.

The cause of pump cavitation may be attributed to one or more of the following conditions:

1. Liquid pumping temperature that is higher than that of the design temperature for the pump, causing higher vapor pressure.
2. The suction head on the pump is lower than the pump manufacturer's recommendations.
3. Suction lift on the pump is higher than the pump manufacturer's recommendations.
4. Running the pump at speeds higher than the pump manufacturer's recommendations.

As flow rate through the pump is increased, the $NPSH_R$ also increases. One way to reduce the $NPSH_R$ is to cut back on the flow rate using a discharge control valve. Of course, this may not be desirable if increasing the flow rate is the objective. In such cases, the available NPSH must be increased by using a small booster pump that itself requires a low NPSH or increasing the suction head on the pump.

Alternatively, using a double suction pump would reduce the NPSH requirement at the higher flow rate.

Summary

In this chapter the NPSH requirement of a pump was reviewed. The difference between the NPSH required by a pump ($NPSH_R$) at a particular flow rate and the NPSH available ($NPSH_A$) in a specific pump and piping configuration were explained and illustrated with examples. Since $NPSH_R$ increases with an increase in pump capacity, it is important to calculate the available NPSH for every possible flow rate scenario and ensure that the absolute pressure at the inlet of the pump suction is always higher than the liquid vapor pressure at the pumping temperature to prevent cavitation of the pump. The available NPSH, being a function of the head loss in the suction piping, always decreases as the flow rate through the pump increases. In pipeline expansion scenarios, care must be exercised to ensure that the main shipping pumps have adequate suction pressure as the system is designed to handle higher flow rates. In many cases, a suction booster pump may have to be installed upstream of the main pumps to prevent pump cavitation.

Problems

7.1 A petroleum liquid is stored in a pressurized tank at 150 psia. The tank is piped to a centrifugal pump located 29 ft away from the tank bottom, similar to the arrangement in Figure 7.3. The liquid level in the tank is 10 ft. The tank bottom and pump suction are at elevations of 22 ft and 25 ft, respectively. The suction piping consists of NPS 10 pipe, 0.250 inch wall thickness, 29 ft long, and the following valves and fittings: two NPS 16 gate valves and four NPS 16 90-degree LR elbows. Calculate the available NPSH at a flow rate of 1800 gpm. If the pump curve shows $NPSH_R$ of 28 ft, will the pump cavitate? Use the Colebrook-White equation with pipe absolute roughness = 0.002 inch and vapor pressure of liquid at pumping temperature = 10.0 psia. Liquid properties are Sg = 0.54 and viscosity = 0.46 cSt.

7.2 An atmospheric gasoline tank supplies product to a centrifugal pump located 10 m away via a DN 250, 6 mm wall thickness pipeline. The total equivalent length of pipes, valves, and fittings from the suction side of pump can be assumed to equal 23 m. The pumps selected for this application requires an NPSH of 8.5 m at a flow rate of 6500 L/min. Is there any danger of cavitation of the pump? What is the value of $NPSH_A$ if the flow rate increases to 7200 L/min?

Chapter 8

Pump Applications and Economics

In this chapter we will review the application of centrifugal pumps and discuss the economic aspects of pumping systems. We will look at pumps in both series and parallel configurations, analyze the combined pump head curves, and explain how they are used in conjunction with system head curves. We will review instances when series pumps are used compared to parallel pumps. The important requirement of matching heads for pumps in parallel will be explained, and the impact of shutting down one or more units in a multiple pump configuration will be analyzed.

Under economics, we will calculate the capital cost and annual operating cost of pumps and pipelines that constitute pumping systems. A cost-benefit analysis is an important exercise that needs to be performed when making investment decisions

DOI: 10.1016/B978-1-85617-828-0.00008-1

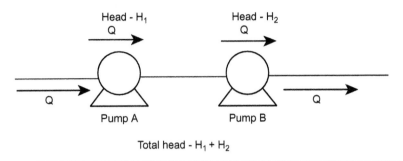

Figure 8.1 Pumps in series.

for replacement of aging equipment as well as when purchasing new facilities to expand pipeline capacities or venture into new markets.

Several case studies of applications of centrifugal pumps and piping systems will be reviewed, tying together the concepts introduced in the previous chapters. In these case studies, we will compare the capital and operating costs of different pumping options and determine the rate of return on capital invested.

Pumps in Series and Parallel

When system head requirements are high, such as due to an increase in pipeline throughput, existing pumps may be replaced with larger pumps. Alternatively, multiple pumps may be installed to produce the additional head required. Multiple pumps afford flexibility, since the system may be run at lower flow rates, if needed, by shutting down one or more pumping units. With one large pump, reducing flow rates means throttling pump head with a discharge control valve, which, as we have seen in earlier chapters, leads to wasted energy and power costs.

When two or more pumps are used in an application, they may be configured in either series or parallel configuration. In a series arrangement, each pump handles the same flow rate, but the total head produced by the combination of pumps will be additive. Since each pump generates a head H corresponding to a flow Q, when configured in series, the total head developed is $H_T = H_1 + H_2$, where H_1, H_2 are the heads developed by the pumps in series at the common flow rate Q. This is illustrated in Figure 8.1, where pump A produces a head H_1 at a capacity of Q, while Pump B produces a head of H_2 at the same capacity Q. Being in series, the combined head is the sum of the two heads.

Figure 8.2 shows the single pump head curve and the combined head curve for two identical pumps in series. Each pump produces a head of $H = 1200$ at a capacity of $Q = 1000\,\text{gpm}$. The combination in series will generate a total head of $H_T = 2400$ at a capacity of $Q = 1000$.

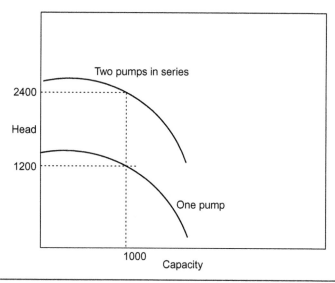

Figure 8.2 Two identical pumps in series.

Suppose there are three identical pumps in series, each producing 1500 ft of head at 1000 gpm capacity. The total head generated by the three pumps in series at 1000 gpm is

$$H_T = 3 \times 1500 = 4500 \text{ ft}$$

If these pumps in series are not identical but instead have differing heads of 1500 ft, 1200 ft, and 1400 ft at 1000 gpm, as shown in Figure 8.3, the total head generated by these pumps in series at Q = 1000 gpm is

$$H_T = 1500 + 1200 + 1400 = 4100 \text{ ft}$$

When two pumps are configured in parallel, the flow rate Q is split between the pumps at the inlet into Q_1 and Q_2, and after passing through the pumps on the discharge side, the flows recombine back to the flow rate of Q, as shown in Figure 8.4. Each pump develops the same head H at the corresponding capacity. Thus, the first pump at capacity Q_1 develops the same head H as the second pump at capacity Q_2. This commonality of head across parallel pumps is the most important feature of pumps installed in parallel. If the pump heads are not matched, pumps in parallel will not function properly.

Consider two identical pumps, each with the H-Q curve, as shown in Figure 8.5. The combined H-Q curve in parallel operation is labeled in the figure as *two pumps in parallel*. At a head of 1200 ft, the capacity of each pump is 1000 gpm. Therefore, in combination, the parallel pumps will be capable of pumping 2000 gpm, generating a common head of 1200 ft. Every point on the combined H-Q curve has a capacity double that of each pump at the same head.

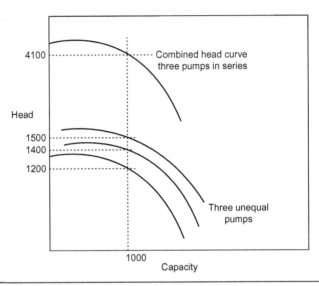

Figure 8.3 Three unequal pumps in series.

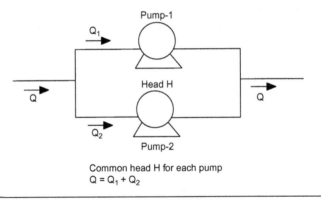

Figure 8.4 Pumps in parallel.

Therefore, when installed in parallel, the flow rates are additive, while the head across each pump is the same. Suppose there are three identical pumps, each developing 800-ft head at a capacity of 400 gpm. When configured in parallel, the flow rate of 1200 gpm is split equally through each pump (400 gpm each), and each pump develops a head of 800 ft. Thus, the total flow is

$$Q_T = 400 + 400 + 400 = 1200 \text{ gpm}$$

And the common head across each pump is $H_1 = H_2 = H_3 = 800$ ft.

To recap, pumps in series have a common flow rate with heads being additive. With pumps in parallel, the flow rates are additive with a common head. The

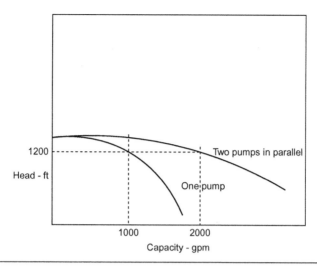

Figure 8.5   Two identical pumps in parallel.

efficiency curve of two or more identical pumps in series or parallel will be the same as each individual pump. However, if the pumps are not identical, we will have to generate a combined E-Q curve from the individual E-Q curves. We will look at some examples of how pump performance in series and parallel are calculated.

EXAMPLE 8.1 USCS UNITS

Two identical pumps with the following data are configured in series.

Q gpm	180	360	420	480
H ft	380	310	278	235
E%	62	82	81	77

Develop the combined pump H-Q curve.

Solution

Since the flow rate is common and heads are additive for pumps in series, we add the H values for each Q value and create the following table of H and Q values for the combined pump performance:

Q gpm	180	360	420	480
H ft	760	620	556	470
E%	62	82	81	77

The efficiency for the combined performance will be the same, since they are identical pumps. The combined pump performance curves are plotted in Figure 8.6.

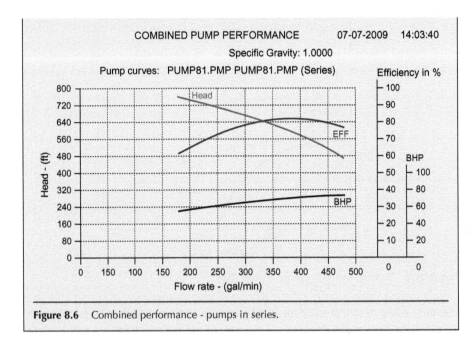

Figure 8.6 Combined performance - pumps in series.

EXAMPLE 8.2 USCS UNITS

Three unequal pumps with the following data are configured in series:

Pump-1

Q gpm	H ft	E%
0.00	1750.00	0.00
500.00	1663.00	63.80
1000.00	1400.00	85.00
1250.00	1203.00	79.70
1500.00	963.00	63.80

Pump-2

Q gpm	H ft	E%
0.00	1225.00	0.00
500.00	1164.00	62.00
1000.00	980.00	83.00
1250.00	842.00	78.00
1500.00	674.00	61.00

Pump-3

Q gpm	H ft	E%
0.00	735.00	0.00
500.00	698.00	60.00
1000.00	588.00	81.00
1250.00	505.00	76.00
1500.00	405.00	59.00

Develop the combined pump head and efficiency curves.

Solution

Since the flow rate is common and heads are additive for pumps in series, we add the H values for each of the three pumps at the same Q value and create the following table of H and Q values for the combined pump performance.

Q gpm	H ft	E%
0.00	3710.00	0.0
500.00	3525.00	62.4
1000.00	2968.00	83.5
1250.00	2550.00	78.4
1500.00	2042.00	61.9

In the combined performance data, the efficiency values in the last column were calculated as follows: Since the pump efficiencies are different at the same flow rate for each pump, the combined efficiency is calculated using Equation (2.5) for BHP. Since the total BHP is the sum of the three BHPs, we have the following equation:

$$Q_T H_T / E_T = Q_1 H_1 / E_1 + Q_2 H_2 / E_2 + Q_3 H_3 / E_3 \qquad (8.1)$$

where subscript T is used for total or combined performance, and 1, 2, and 3 are for each of the three pumps.

Simplifying Equation (8.1), by setting $Q_1 = Q_2 = Q_3 = Q_T$ for series pumps and solving for the combined efficiency, E_T we get

$$E_T = H_T / (H_1 / E_1 + H_2 / E_2 + H_3 / E_3) \qquad (8.2)$$

To illustrate how E_T is calculated, we will substitute the values corresponding to $Q_1 = 500$ in Equation (8.2):

$$E_T = 3525/((1663/63.8) + (1164/62.0) + (698/60)) = 62.4\%$$

at $Q = 500$ gpm. Similarly, other values of the combined efficiency are calculated for the remaining values of Q.

EXAMPLE 8.3 SI UNITS

Two identical pumps with the data in the following table are configured in parallel. Determine the combined pump head and efficiency curves.

Q L/min	1200	2160	2520	2880
H m	190	155	139	118
E%	62	82	81	77

Solution

For pumps in parallel, the flow rate is additive for the common value of head. Since these are identical pumps, each pump will handle half the total flow at a common head. At a head of 190 m, each pump has a capacity of 1200 L/min. Therefore, a combined flow of 2400 L/min can be produced by these two parallel pumps at a head of 190 m. Similarly, the common head of 155 m will produce a total flow of 4320 L/min, and so on.

The combined performance is tabulated as follows:

Q L/min	2400	4320	5040	5760
H m	190	155	139	118
E%	62	82	81	77

Since the pumps are identical, the combined efficiencies will be the same as the original pumps for each of the data points. The combined pump performance curves are plotted in Figure 8.7.

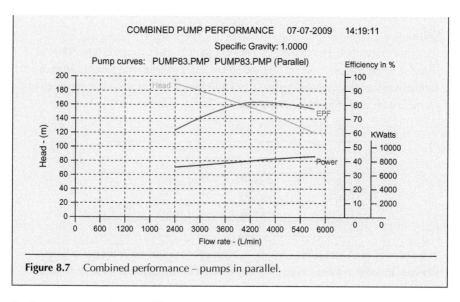

Figure 8.7   Combined performance – pumps in parallel.

In the next example we will examine the combined performance of two unequal pumps in parallel configuration.

EXAMPLE 8.4 SI UNITS

Two unequal pumps with the following performance data are configured in parallel. Determine the combined pump performance.

Pump-1

Q m³/h	H m	E%
0.00	250.00	0.00
100.00	240.00	63.80
150.00	190.00	85.00
175.00	140.00	79.70
200.00	100.00	63.80

Pump-2

Q m³/h	H m	E%
0.00	250.00	0.00
80.00	240.00	60.00
160.00	190.00	79.00
240.00	140.00	81.00
280.00	100.00	76.00

Solution

For pumps in parallel, we add the flow rates of each pump at a common head. Thus, at a head of 240 m, the flow rates of 100 and 80 are added to get a total flow of 180 m³/h. Similarly, at the other common heads of 190, 140, and so on, the flow rates are added to get the combined performance as tabulated next.

Combined Pump Performance

Q m³/h	H m	E%
0.00	250.00	0.00
180.00	240.00	62.05
310.00	190.00	81.79
415.00	140.00	80.45
480.00	100.00	76.00

In the table of combined pump performance, the efficiency values in the last column were calculated as follows: Since the pump efficiencies are different at the same head for each pump, the combined efficiency is calculated using Equation (2.6) for power. Since the total power is the sum of the two powers, we have the following equation:

$$Q_T H_T / E_T = Q_1 H_1 / E_1 + Q_2 H_2 / E_2 \qquad (8.3)$$

where subscript T is used for combined performance, and 1, 2, are for each of the two pumps.

Simplifying Equation (8.3), by setting $H_1 = H_2 = H_T$ for parallel pumps and solving for the combined efficiency E_T, we get

$$E_T = Q_T / (Q_1 / E_1 + Q_2 / E_2) \qquad (8.4)$$

To illustrate how E_T is calculated, we will substitute the values corresponding to $H_1 = 240$ in Equation (8.4):

$$E_T = 180 / ((100/63.8) + (80/60.0)) = 62.05\%$$

at H= 240 m. Similarly, other values of the combined efficiency are calculated for the remaining values of H.

EXAMPLE 8.5 USCS UNITS

Can the following pumps be operated in parallel? If so, what range of flow rates and pressures are possible?

Pump-A

Q gpm	200	360	420	480
H ft	1900	1550	1390	1180
E%	62	82	81	77

Pump-B

Q gpm	200	360	420	480
H ft	2200	2100	1800	1250
E%	63	85	80	75

Solution

For pumps in parallel, there must be a common head range. By inspection, it can be seen that both pumps have a common head range as follows:

$$\text{Minimum head} = 1250\,\text{ft} \quad \text{Maximum head} = 1900\,\text{ft}$$

Corresponding to the preceding head range, the capacity range for the pumps is as follows:

$H = 1250\,\text{ft}$
Pump-A: Q is between 420 and 480 gpm
Pump-B: $Q = 480$ gpm
$H = 1900\,\text{ft}$
Pump-A: $Q = 200$ gpm
Pump-B: Q is between 360 and 420 gpm

Therefore, these two pumps in parallel will be suitable for an application with the following range of flow rates and pressures:

$$H = 1250\,\text{ft} \quad Q_T = 900 \text{ to } 960\,\text{gpm}$$
$$H = 1900\,\text{ft} \quad Q_T = 560 \text{ to } 620\,\text{gpm}$$

Using interpolated Q values, the range of capacity and heads are

$$Q = 590 \text{ to } 930\,\text{gpm and } H = 1900 \text{ to } 1250\,\text{ft}$$

It can be seen that this is quite a narrow range of head and capacity values for these two pumps in parallel. The actual flow rate that can be obtained with these pumps in parallel will depend on the system head curve and will be illustrated using an example.

EXAMPLE 8.6 SI UNITS

Can the following two pumps be operated in series? If so, what range of flow rates and pressures are possible?

Pump-A

Q L/min	2000	3600	4200	4800
H m	190	155	139	118
E%	62	82	81	77

Pump-B

Q L/min	3000	5400	6300	7200
H m	220	210	180	125
E%	63	85	80	75

Solution

For pumps in series, there must be a common capacity range. Reviewing the data provided, it can be seen that both pumps have a common range of capacity as follows:

Minimum capacity = 3000 L/min   Maximum capacity = 4800 L/min

Corresponding to the preceding capacity, the range of heads for the two pumps is as follows:

Q = 3000 L/min
Pump-A: H is between 155 and 190 m
Pump-B: H = 220 m
Q = 4800 L/min
Pump-A: H = 118 m
Pump-B: H is between 210 and 220 m

Therefore, these two pumps in series will be suitable for an application with following range of flow rates and pressures:

$$Q = 3000\,L/min\ H_T = 375\ to\ 410\,m$$
$$Q = 4800\,L/min\ H_T = 328\ to\ 338\,m$$

Using interpolated H values, the range of capacity and heads are

$$Q = 3000\ to\ 4800\,L/min\ and\ H = 393\ to\ 333\,m$$

It can be seen that this is quite a narrow range of head and capacity values for these two pumps in series. The actual flow rate that can be obtained with these pumps in series will depend on the system head curve and will be illustrated in Example 8.7.

EXAMPLE 8.7 USCS UNITS

Two identical pumps, each with the following head/capacity/efficiency data, are configured in series, at a pump station.

Q gpm	1000	2000	3000	4000	5000
H ft	193	184	172	155	132
E%	54	66	75.5	80	78.5

Water (Sg = 1.0) is pumped from a storage tank at Denver located at a distance of 20 ft from the pumps to another tank located 15,700 ft away at Hampton, as shown in Figure 8.8.

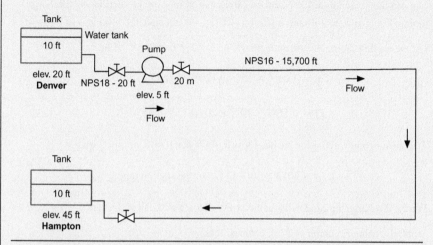

Figure 8.8   Pumping from Denver to Hampton.

The interconnecting pipeline from the discharge of the pumps at Denver is NPS 16, 0.250 inch wall thickness. The suction piping is NPS 18, 0.250 inch wall thickness. Additional data on the elevation of the tanks and pumps are as follows:

Elevations above MSL:

Denver tank bottom: 20 ft Hampton tank bottom: 45 ft
Centerline of pump suction: 5 ft

Use the Hazen-Williams formula with a C factor of 120.

(a) Consider the initial level of water in both tanks as 10 ft, develop the system head curve, and determine the initial flow rate with the two pumps running in series.

(b) If one pump shuts down, what is the resultant flow rate?

To account for valves and fittings, assume 5 psi minor losses on the suction side of the pumps and a loss of 10 psi on the discharge piping.

Solution

Using the elevations given, first calculate the suction head and discharge head on the pumps:

Suction head = 20 + 10 − 5 = 25 ft
Discharge head = 45 + 10 − 5 = 50 ft
Total static head = 50 − 25 = 25 ft

Next, we will calculate the frictional head loss at any flow rate Q. The head loss at Q gpm will be due to the 20 ft of 18-inch suction pipe; 15,700 ft of 16-inch discharge piping; and the given minor losses due to valves and fittings on the suction and discharge piping. Using Hazen-Williams equation (4.27a), we calculate the head loss as follows:

For the suction piping, inside diameter D = 18 − 2 × 0.250 = 17.5 in.

$$h = 1.0461 \times 10^4 \ (Q/120)^{1.852} \ (1/17.5)^{4.87}$$

$$Q = 6.7547 \times 10^{-3} \ (120) \ (17.5)^{2.63} \ (h_s)^{0.54}$$

Transposing and solving for the head loss h_s (in ft per 1000 ft of pipe), we get

$$h_s = 1.3013 \times 10^{-6} \ Q^{1.85} \text{ for the suction piping}$$

For the discharge piping, inside diameter D = 16 − 2 × 0.250 = 15.5 in.

Similarly, using Equation (4.27a), for head loss

$$Q = 6.7547 \times 10^{-3} \ (120) \ (15.5)^{2.63} \ (h_d)^{0.54}$$

$$h_d = 2.3502 \times 10^{-6} \ Q^{1.85} \text{ for the discharge piping}$$

where h_s and h_d are the head loss in the suction and discharge piping in ft of water per 1000 ft of pipe at a flow rate of Q gpm.

The total head loss per 1000 ft is then equal to $h_s + h_d$

$$h_s + h_d = 1.3013 \times 10^{-6} \ Q^{1.85} + 2.3502 \times 10^{-6} \ Q^{1.85}$$
$$= 3.6515 \times 10^{-6} \ Q^{1.85}$$

Multiplying by the pipe length, the total head loss H_s is

$$(15,700/1000) \times 3.6515 \times 10^{-6} \ Q^{1.85} = 5.7329 \times 10^{-5} \ Q^{1.85} \text{ ft}$$

To this we must add the minor losses due to the suction and discharge piping. The system head H_s is therefore obtained by adding the minor losses and 25 ft static head calculated earlier:

$$H_s = 5.7329 \times 10^{-5} \ Q^{1.85} + (5 + 10) \times 2.31/1.0 + 25$$

$$H_s = 5.7329 \times 10^{-5} \ Q^{1.85} + 59.65 \text{ ft} \qquad (8.5)$$

where H_s is the system head required at any flow rate Q gpm.

We next develop the system head curve from Equation (8.5) for a set of flow rates from 1000 to 5000 gpm, as follows:

System Head Curve

Q gpm	1000	2000	4000	5000
H_s ft	80.01	133.06	324.28	459.52

This system head curve is plotted with the combined pump H-Q curve of the given two identical pumps in series, as shown in Figure 8.9. The single pump head curve is also shown.

(a) The initial flow rate with the two pumps running in series is indicated by the point of intersection of the system head curve and the combined pump head curve labeled *Two pumps* in Figure 8.9. This is approximately 3900 gpm.
(b) If one pump shuts down, the resultant flow rate is indicated by the point of intersection of the system head curve and the *single* pump head curve labeled *one pump* in Figure 8.9. This is approximately 2600 gpm.

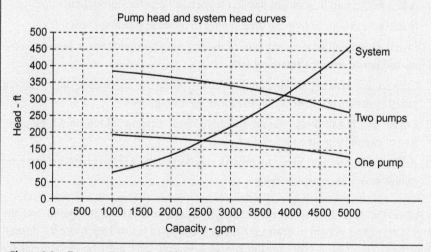

Figure 8.9   Two pumps in series with system head curve.

Economics of Pumping Systems

Frequently, a system of pumping equipment has been in operation for 20 years or more and may have performed satisfactorily for several years after initial installation. However, the operational efficiency may have degraded due to wear and tear of the equipment, resulting in a drop in pump efficiency. Although regular maintenance may have been performed on such equipment, efficiencies can never reach the original values at the time of initial installation. Therefore, modification or replacement of pumps and ancillary equipment may be necessary due to such reasons, as well as due to changes in pumping requirements triggered by business conditions. Alternatively, due to changes in market conditions, the pumping system in recent years may be operating well below the originally designed optimum capacity when demands were higher. This would cause operation of facilities at less than optimum conditions. Such situations also require examination of equipment to determine whether modifications or replacements are necessary. In all such cases, we need to determine the additional capital cost and annual operating costs of the modifications, and then compare these costs with the benefits associated with increased revenue due to added equipment.

Here are some economic scenarios pertaining to pumps and pump stations:

1. Change existing pump impeller and install larger motor if necessary to handle additional pump capacity required by a new business.
2. Add a new pump in series or parallel to increase pipeline throughput.
3. Install new pump, or loop existing piping to increase pipeline capacity.
4. Install a crude oil heater upstream of the shipping pump to reduce liquid viscosity and hence enhance pumping rates.
5. Convert an existing turbine/engine-driven pump to an electric motor-driven pump (constant speed or VFD) to reduce operating costs.
6. Locate and install a suitable intermediate booster pump station on a long pipeline to expand pipeline throughput.
7. Add a tank booster pump to provide additional NPSH to mainline shipping pumps and to increase capacity.

All of the preceding scenarios require hydraulic analysis of the pump and the associated piping system to determine the best course of action to achieve the desired objectives. The capital cost and the annual operating cost must be determined for each option so that a sound decision can be made after examining the cost versus benefits. We will examine case studies of some of the preceding scenarios next.

1. Change existing pump impeller and install larger motor if necessary to handle additional pump capacity.
In this case, a pump may have been fitted with a 15-inch impeller to satisfy the pumping requirements in place five years ago. Recently, due to additional market demand, an incremental pumping requirement has been proposed. The least-cost alternative is probably one in which additional pump capacity may be obtained by changing

the pump impeller to the maximum diameter impeller possible within the pump casing. Assuming that a 17-inch impeller can be installed in this pump, we need to determine if the larger impeller will demand more power than the existing drive motor can provide. It may be necessary to replace the existing 2000-HP motor with a 2500- or 3000-HP motor. This case will require analysis of the additional cost of pump impeller modification, as well as the cost of purchasing and installing a larger electric motor drive.

EXAMPLE 8.8 USCS UNITS

The Ajax pipeline company has been shipping petroleum products from their Danby refinery to a distribution terminal located 54 miles away in the town of Palmdale. The existing pumping system installed ten years ago consists of a centrifugal pump Model 6 × 8 × 13 MSN (10-inch impeller) as listed.

Q gpm	0	1050	2100	2625	3150
H ft	1150	1090	1000	900	600
E%	0.0	63.8	85.0	79.7	63.8

The pump takes suction from an elevated tank located 32 ft away. The suction and discharge piping are as shown in Figure 8.10.

The pipeline from Danby to Palmdale consists of 54 miles of NPS 16, API 5LX-52 pipe with a wall thickness of 0.250 in. The maximum allowable operating pressure (MAOP) of the pipeline is limited to 1150 psig. The throughput in recent years has been ranging from 2600 to 3000 bbl/h of mostly gasoline (Sg = 0.74 and visc = 0.6 cSt at 60°F).

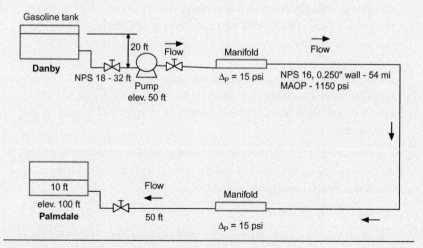

Figure 8.10   Danby to Palmdale gasoline pipeline.

Recently, *additional* market demand of 860,000 bbl per month has been identified. The least-cost alternative of installing the maximum impeller diameter (13 in.) in the existing pump along with a larger motor drive has been proposed. Another alternative includes the use of drag reduction additive (DRA), which may possibly require less investment but additional operating costs.

(a) Determine the capital cost and increased operating cost associated with the larger pump impeller and electric motor replacement.

(b) If the current transportation tariff is $0.05 per bbl, estimate the rate of return on this project. Assume an interest rate of 5% and a project life of 15 years. Use the Hazen-Williams equation with $C = 140$.

Solution

Our approach to this problem will be as follows:

1. Using the given pump curve data, confirm the operating point (3000 bbl/h) by developing a system head curve and plotting the pump head curve to determine the point of intersection.

2. Determine the new flow rate corresponding to the increased demand of 860,000 bbl/month = $860000/(30 \times 24)$ = 1195 bbl/h.
   New flow rate = $3000 + 1195 = 4195$ bbl/h

3. From the system head curve determine the pump pressure required at the new operating point corresponding to 4195 bbl/h.

4. Using Affinity Laws, develop a new pump head curve for a 13-inch impeller based on the current 10-inch impeller.

5. Confirm that the 13-inch impeller can provide the necessary head at the higher flow rate identified by the system head curve.

6. Determine increased drive motor HP requirement at 4195 bbl/h.

7. Calculate incremental capital cost for installing larger impeller and for a new electric motor.

8. Calculate incremental operating cost at the higher flow rate.

9. Calculate increased revenue from given tariff rate.

10. Determine the rate of return (ROR) considering a project life of 15 years and an interest rate of 5%.

At the present flow rate of 3000 bbl/h, we will determine the pressure drop in the suction and discharge piping and calculate the system head required. In the expansion scenario an 860,000 bbl/mo increase is expected.

Final flow rate = $3000 + 860,000/(30 \times 24) = 4195$ bbl/h

The process of system head calculation will be repeated for a range of flow rates from 2000 bbl/h to 4195 bbl/h to develop the system head curve.

For flow rate Q = 3000 bbl/h (2100 gpm), using Equation (4.27) for Hazen-Williams with C = 140, we calculate the head loss in the NPS 16 discharge piping as h_{fd} = 2.52 ft/1000 ft. Similarly, for the NPS 18 suction line h_{fs} = 1.39 ft/1000 ft. We could also use the tables in Appendix G to calculate the head losses.

The system head is calculated by taking into account the suction head, discharge head, the head losses in the suction and discharge piping, and the meter manifold losses.

$$H_s = (50 - 20 + 1.39 \times 32/1000 + 2.52 \times 54 \times 5280/1000) \text{ ft} + (30 + 50) \text{ psi}$$

$$H_s = 748.55 + 80 \times 2.31/0.74 = 998.28 \text{ ft at } Q = 3000 \text{ bbl/h (2100 gpm)}$$

Similarly, H_s is calculated for flow rates of 1000, 2100, 2500, 3000, and 4000 gpm. These correspond to a range of 1429 bbl/h to 5714 bbl/h. The following table is prepared for the system head curve:

Q gpm	1000	2100	2500	3000	4000
H_s ft	461	998	1271	1669	2646

It can be seen that at the initial flow rate of 3000 bbl/h (2100 gpm), the pump produces 1000 ft of head, which is close to the 998 ft required by the system head curve. This is therefore the initial operating point.

Plotting the system head curve shows the head required (1615 ft) at the final flow rate of 4195 bbl/h (approx. 3000 gpm), as shown in Figure 8.11. Obviously, this operating point is unattainable using the existing 10 in. impeller. Hence, we consider the use of the maximum diameter impeller of 13 in.

Using Affinity Laws, the pump head curve for the 13 in. impeller is as follows:

Q gpm	0	1365	2730	3412.5	4095
H ft	1944	1842	1690	1521	1014

Plotting this 13 in. diameter pump curve on the system head curve, it shows the point of operation to be at 3000 gpm or 4286 bbl/h, as indicated in Figure 8.12. This is close to the final flow rate of 4195 bbl/h desired. Therefore, the 13 in. impeller, slightly trimmed, can provide the necessary head at the higher flow rate. In fact, we can prove that the impeller size required is actually 12.88 inch for 4195 bbl/h flow rate.

(a) The cost of the new impeller (13 in.) and the rotating assembly will be around $50,000. Installation and testing will be an additional $25,000, for a total of $75,000.

At the higher flow rate, more power is required with the 13 in. impeller. Therefore, the existing electric motor will be replaced with a higher-horsepower electric motor.

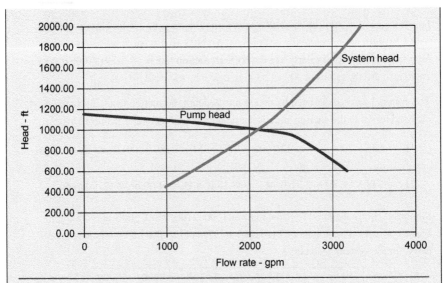

Figure 8.11 System head curve and pump head curve for 10" impeller.

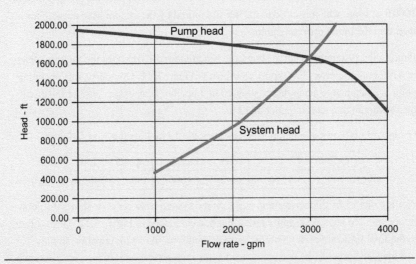

Figure 8.12 System head curve and pump head curve for 13" impeller.

This motor replacement to handle the higher flow rate will cost $150,000, including labor and materials. Therefore, the total capital cost for the pump impeller and motor is $225,000.

The operating cost will increase due to the additional power consumption at the higher flow rate. Using Equation (2.5):

$$BHP = (4195 \times 0.7) \times 1615 \times 0.74/(3960 \times 0.84) = 1055$$

Considering 95% motor efficiency, motor HP $= 1055/0.95 = 1111$.

We must also check power required at the maximum pump capacity. This is found to be at Q $= 4057$ gpm, H $= 995$ ft at E $= 63.8\%$.

$$BHP = 4057 \times 995 \times 0.74/(3960 \times 0.638) = 1183$$

Considering 95% motor efficiency, motor HP $= 1183/0.95 = 1246$

The nearest standard size motor is 1250 HP.

Electrical energy consumption per day

$$= 1111 \times 0.746 \times 24 = 19{,}892 \text{ kWh/day}$$

At $0.10/kWh, annual power cost $= 19{,}892 \times 365 \times 0.1 = \$726{,}058/yr$

Prior to the increase in pipeline throughput the power consumption at initial flow rate of 3000 bbl/h is

$$BHP = (3000 \times 0.7) \times 1000 \times 0.74/(3960 \times 0.85) = 462$$
Power cost $= (462/0.95) \times 0.746 \times 24 \times 365 \times 0.1 = \$317{,}806$

Therefore, the increase in power cost is

$$= \$726{,}058 - \$317{,}806 = \$408{,}252$$

(b) The revenue increase at $0.05 per bbl is

$$(4195 - 3000) \times 0.05 \times 365 \times 24 = \$523{,}410/yr$$

For a capital investment of $225,000 and a net yearly revenue ($523,410 – 408,252) of $115,158 after deducting the power cost, we get a rate of return of 51.08% using a discounted cash flow (DCF) method (project life 15 years).

2. Add a new pump in series or parallel to increase pipeline throughput.

In this scenario an existing installation has been in operation for a while, with a single pump providing the necessary pipeline throughput. As before, increased demand requires analysis of possible capacity expansion of the pumping system.

One alternative is to add a new pump in series or parallel to the existing pump, thereby increasing the pumping rate as described in the next example.

Other options include the use of DRA to reduce friction in the pipeline and thereby increase throughput without additional pump pressure. This option will reduce pumping power at a higher flow rate, but power requirements will increase. Hence, a new drive motor will be required at the higher flow rate, even if the pump head curve has the additional capacity.

EXAMPLE 8.9 SI UNITS

The Andes Pipeline Company operates a 24 km crude oil pipeline system that transports light crude (Sg = 0.85 and visc = 10 cSt at 20°C) from Bogota (elevation 1150 m) to Lima (elevation 1270 m). Increased market demand necessitates expanding pipeline capacity from the present $1000\,m^3/h$ to $1200\,m^3/h$. Existing equipment at Bogota pump station includes one centrifugal pump with the following data:

Q m³/h	0	500	1000	1250	1500
H m	632.5	600.9	506.0	434.8	347.9
E%	0.0	60.0	80.0	75.0	60.0

The pump is connected to the crude oil tanks as shown in Figure 8.13. Delivery pressure required at Lima = 3.5 bar. Pressure losses in meter manifolds at Bogota and Lima are 1.2 bar each.

It has been proposed to increase pipeline throughput by installing a second identical pump in series or parallel as necessary. Compare this alternative with using DRA. The latter option requires installation of a skid-mounted DRA injection system that can be leased annually at $60,000. The DRA is expected to cost $5 per liter delivered to the site. It is also estimated that DRA injection rate of 10 ppm will be required to achieve the increase in flow rate from $1000\,m^3/h$ to $1200\,m^3/h$.

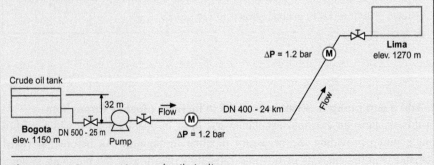

Figure 8.13   Bogota to Lima crude oil pipeline.

Solution

Our approach to this problem will be as follows:

1. Using the given pump curve data, confirm the operating point ($1000\,m^3/h$) by developing a system head curve and plotting the pump head curve to determine the point of intersection.
2. Determine from the system head curve the new operating point for the increased flow rate scenario ($1200\,m^3/h$).
3. Select a suitable two pump configuration (series or parallel) to handle the increased flow rate. Determine necessary impeller trim.
4. Calculate operating cost for initial and final flow rate scenarios.
5. Calculate incremental capital cost for second pump.
6. Calculate capital and operating cost for DRA option.
7. Compare the second pump scenario with the DRA option for capital and operating cost.

At the present flow rate of $1000\,m^3/h$ we will determine the pressure drop in the suction and discharge piping and calculate the system head required. The process of system head calculation will be repeated for a range of flow rates from $500\,m^3/h$ to $2000\,m^3/h$ to develop the system head curve.

Suction piping:

Inside diameter $D = 500 - 2 \times 10 = 480\,mm$

At $Q = 1000\,m^3/h$, using Equation (4.10), the Reynolds number is

$$R = 353678 \times 1000/(480 \times 10) = 73{,}683$$

Relative roughness $e/D = 0.05/480 = 0.0001$

$$e/(3.7D) = 0.0001/3.7 = 2.8153 \times 10^{-5}$$

From Colebrook-White equation (4.21), the friction factor f is

$$1/\sqrt{f} = -2\,Log_{10}[2.8153 \times 10^{-5} + 2.51/(73683\sqrt{f})]$$

Solving for f by trial and error:

$$f = 0.0197$$

The head loss in the suction piping, from Equation (4.14) is

$$P_{km} = 6.2475 \times 10^{10} \times 0.0197 \times 1000^2 \times 0.85/480^5 = 41.0568\ kPa/km$$

Converting to head loss in m/km using Equation (1.12):

$$h_{fs} = 41.0568 \times 0.102/0.85 = 4.93\ m/km$$

Similarly, we calculate the head loss for the discharge piping next.

Discharge piping:

Inside diameter $D = 400 - 2 \times 8 = 384\,mm$

At $Q = 1000\,m^3/h$, using Equation (4.10), the Reynolds number is

$$R = 353678 \times 1000/(384 \times 10) = 92{,}104$$

Relative roughness $e/D = 0.05/384 = 0.00013$

$$e/(3.7D) = 0.00013/3.7 = 3.5135 \times 10^{-5}$$

From Colebrook-White equation (4.21) the friction factor f is

$$1/\sqrt{f} = -2\,Log_{10}[3.5135 \times 10^{-5} + 2.51/(92104\sqrt{f})]$$

Solving for f by trial and error:

$$f = 0.0189$$

The head loss in the discharge piping, from Equation (4.14) is

$$P_{km} = 6.2475 \times 10^{10} \times 0.0189 \times 1000^2 \times 0.85/384^5 = 120.21\ kPa/km$$

Converting to head loss in m/km using Equation (1.12):

$$h_{fd} = 120.21 \times 0.102/0.85 = 14.43\ m/km$$

Therefore, the system head at $Q = 1000\,m^3/h$ can be calculated by adding the suction and discharge piping head losses to the net discharge head and meter manifold losses of 2.4 bar and delivery pressure of 3.5 bar as follows:

$H_s = (14.43 \times 24)\ m + (4.93 \times 0.025)\ m + (1270 - 1150 - 32)\ m + (2.4 + 3.5)\ bar$
$H_s = 434.44 + 590 \times 0.102/0.85 = 505.24\ m$

Therefore, $Q = 1000\,m^3/h$ $H_s = 505.24\,m$.

Similarly, the system head is calculated at flow rates of 500, 1200, 1500, and $2000\,m^3/h$, and the following table is prepared for the system head curve:

Q m³/h	500	1000	1200	1500	2000
H_s m	259.0	505.24	643.0	886.3	1393.0

The system head curve is plotted along with the pump head curve as shown in Figure 8.14. It can be seen that the initial flow rate of $1000\,m^3/h$ is attainable with the existing pump.

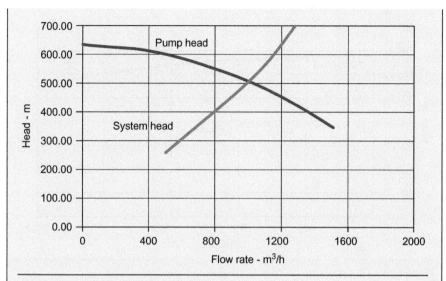

Figure 8.14   Initial flow rate – 1000 m³/h One pump.

Examining the system head curve in Figure 8.14, we see that at the increased flow rate scenario of 1200 m³/h, the head required is 643 m. In order to obtain this head we will consider the use of two identical pumps in series or parallel.

Series pump option:

In this case, two identical pumps with the current impeller size will produce approximately 900 m of head. Since our requirement is 643 m, extensive impeller trimming will be needed and is not advisable.

Parallel pump option:

In this case, we install two identical pumps in parallel such that in combination they produce a total capacity of 1200 m³/h at a head of 643 m. This will require increasing the impeller size to approximately 105% of the present impeller size. Assuming the existing impeller is 250 mm diameter, we will require an impeller size of approximately 262 mm in each pump. The combined pump curve is shown plotted on the system head curve in Figure 8.15.

The operating cost for the initial and final flow rates is calculated next. Initial power cost at 1000 m³/h using the existing pump is calculated from Equation (2.4):

Power = $1000 \times 506 \times 0.85/(367.46 \times 0.8) = 1463$ kW

Considering a motor efficiency of 95%, the motor power required = $1463/0.95 = 1540$ kW.

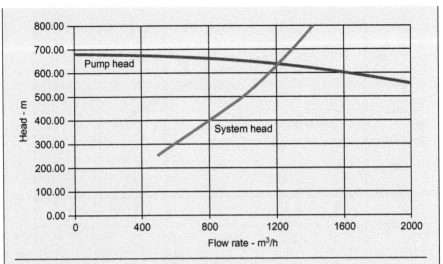

Figure 8.15 Final flow rate – 1200 m³/h Two pumps in parallel

Final power cost at 1200 m³/h using two parallel pumps is calculated from Equation (2.4):

Power = 600 × 645 × 0.85/(367.46 × 0.6996) = 1280 kW for each pump
The motor power required = 2 × 1280/0.95 = 2695 kW
Increased power required from 1000 m³/h to 1200 m³/h = 2695 – 1540 = 1155 kW
At $0.10/kWh increased energy cost = 1155 × 24 × 365 × 0.1 = $1,011,780/yr

The capital cost of adding a pump in parallel will consist of

1. Trimmed impeller for original pump: $75,000
2. Second pump and motor: $250,000
3. Additional piping to accommodate second pump: $50,000
4. Installation cost: $100,000

Thus, the total capital cost for the two pump options will be approximately $475,000. We will examine the DRA option next.

The DRA skid will be leased at $60,000/yr, and the DRA will be purchased at a cost of $5/L. Considering a DRA injection rate of 10 ppm at the pipeline flow rate 1200 m³/h, the DRA option is as follows.

DRA required per year:
= (10/10⁶) × 1200 × 24 × 365 = 105.12 m³/yr = 105,120 L/yr
DRA cost per year = 5 × 105,120 = $525,600 per year
Total cost of DRA option per year = $525,600 + $60,000 = $585,600

The DRA option will require changing out the existing electric motor drive, since at the higher flow rate the existing pump will require additional power. Therefore, the DRA option will require a capital investment of approximately $150,000.

In addition, the DRA option will result in increased power consumption at the increased flow rate of 1200 m³/h. This can be calculated from the original pump curve as follows:

Power required at 1200 m³/h = 1200 × 451 × 0.85/(367.46 × 0.769) = 1628 kW
Considering a motor efficiency of 95%, the motor power required
   = 1628/0.95 = 1714 kW
Increased power required from 1000 m³/h to 1200 m³h = 1714 – 1540 = 174 kW

At $0.10/kWh, the increased power cost = 174 × 24 × 365 × 0.1 = $152,424/yr

Thus, the DRA option is as follows:

Capital investment = $150,000.
Total annual cost for DRA and power
   = $585,600 + 152,424 = $ 738,024

The two options of increasing capacity from 1000 m³/h to 1200 m³/h are summarized as follows:

	Two-Pump Option	DRA Option
Capital Cost	$475,000	$150,000
Annual Cost	$1,011,780	$738,024

It can be seen that the DRA option is more cost effective and hence is the better of the two options.

3. Install new pipe or loop existing piping to increase pipeline capacity.

A pipeline and pumping system is originally designed for some initial capacity, with room for some amount of expansion in throughput without exceeding the liquid velocity limits. Suppose initially a capacity of 5000 gpm resulted in an average flow velocity of 6 ft/s and corresponding pressure drop of 15 psi/mi of pipe length. If the system were to be expanded to a capacity of 8000 gpm, the average velocity in the pipe will increase to approximately (8000/5000) × 6 = 9.6 ft/s. This increase in velocity will result in a pressure drop of 2.5 times the original pressure drop. In some cases, such an increase in velocity and pressure drop may be tolerable. When such high velocities and pressure drops cannot be allowed, we have to resort to installing a larger-diameter pipeline to handle the increased flow rate. The cost of this will be prohibitively expensive. As an option, we could consider looping a section of the pipeline by installing a pipe of the same diameter in parallel. This will effectively increase the flow area and result in lower velocity and lower pressure drop. Generally, looping large sections of an existing pipeline may not be very cost effective. Nevertheless, we must compare this option with that of adding pumps at the origin of the pipeline to increase throughput, as discussed in the Example 8.10.

EXAMPLE 8.10 USCS UNITS

The Ajax pipeline in Example 8.8 needs to be expanded to increase capacity from the existing 3000 bbl/h to 4000 bbl/h (2100 gpm to 2800 gpm) of gasoline.

(a) Starting with the existing pump (10 in. impeller) at Danby, how much of the 54-mi pipeline should be looped to achieve the increased capacity?

(b) Compare the cost of looping the pipe versus installing a single new pump at Danby. The existing pump will remain as a spare unit. Assume $1500/ton for pipe material cost and $50/ft for pipe installation.

Solution

Our approach to this problem will be as follows:

1. Using the given pump curve data, confirm the operating point of 3000 bbl/h (2100 gpm) by developing a system head curve and plotting the pump head curve to determine the point of intersection. This was done in Example 8.8.

2. From the system head curve, determine the pump head required at the new operating point C in Figure 8.16, corresponding to the increased flow rate of 4000 bbl/h (2800 gpm).

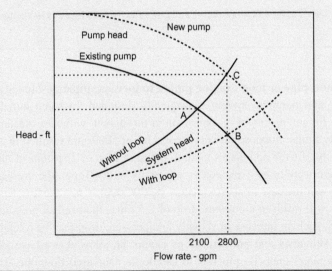

Figure 8.16 Ajax pipeline system head and pump head curves.

3. For the new pump option, select a suitable pump to handle the operating point C on the original system head curve, as in Figure 8.16.

4. On the existing pump head curve, determine the head available at point B corresponding to the new flow rate of 2800 gpm (4000 bbl/h).

5. Determine by trial and error the length of pipe loop required to create a lower system head curve that will result in the operating point B at the new flow rate.

6. Calculate the capital and operating costs of the new pump and motor option.

7. Calculate the capital and operating costs of the looped pipe option.

8. Calculate incremental operating costs at the higher flow rate for both options.

9. Compare the rate of return for the two options.

In Example 8.8, the following system head curve was generated:

Q gpm	1000	2100	2500	3000	4000
H_s ft	461	998	1271	1669	2646

From the plot of the system head curve, we find that at $Q = 2800$ gpm (4000 bbl/h), the system head $H_s = 1503$ ft.

New pump option:

We will select a new pump to satisfy the preceding condition. The following pump curve is proposed for the application:

Q gpm	0	1400	2800	3500	4200
H_s ft	1879	1785	1503	1292	1033
E%	0	60	80	75	60

This new pump is capable of handling the increased flow rate of 4000 bbl/h (2800 gpm) as shown in Figure 8.17. The power required is calculated using Equation (2.5) as follows:

$$BHP = 2800 \times 1503 \times 0.74/(3960 \times 0.8) = 983$$

Using a motor efficiency of 0.95, the motor HP required = 983/0.95 = 1035. The nearest standard size motor of 1250 HP is selected.

Looping option:

In this option, we use the original pump that was used at the initial flow rate of 3000 bbl/h and determine the length of the 54-mi pipeline section to be looped. Initially, assume that a 20-mi section of the pipeline is looped with an identical NPS 16 pipe. From Equation (4.42a), we determine the equivalent diameter of the two NPS 16 pipes as

$$D_E = 15.5 \, (2)^{0.4} = 20.45 \text{ in.}$$

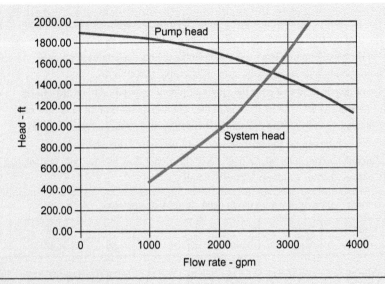

Figure 8.17 New pump and system curves for 4000 bbl/hr.

Thus, we have a pipeline consisting of 20 mi of 20.45 in. inside-diameter pipe (representing the looped section) and 34 mi of NPS 16 pipe from Danby to Palmdale. For this pipeline, a new system head curve will be developed for the range of Q values from 1000 to 4000 gpm, as described next.

For the flow rate $Q = 2100$ gpm, using Equation (4.27) for Hazen-Williams with $C = 140$, we calculate the head loss in the NPS 16 discharge piping as

$$h_{fd} = 2.52 \text{ ft/1000 ft}$$

Similarly, for the NPS 18 suction line, it is

$$h_{fs} = 1.39 \text{ ft/1000 ft}$$

and for the equivalent diameter $D = 20.45$ in., the head loss is

$$h_{fe} = 0.65 \text{ ft/1000 ft}$$

Note that these head losses can also be calculated using the Appendix G tables and adjusting the table values for $C = 140$.

The system head is calculated by taking into account the suction head, the discharge head, the head losses in the suction, the discharge piping, and the meter manifold losses.

$$H_s = (50 - 20 + 1.39 \times 32/1000 + 2.52 \times 34 \times 5280/1000$$
$$+ 0.65 \times 20 \times 5280/1000)\,\text{ft} + (30 + 50)\,\text{psi}$$

$$H_s = 551.08 + 80 \times 2.31/0.74 = 801\,\text{ft at } Q = 2100\,\text{gpm}$$

Similarly, H_s is calculated for flow rates of 1000, 2100, 2500, 3000, and 4000 gpm. The following table is prepared for the system head curve for the looping option:

Q gpm	1000	2100	2500	3000	4000
H_s ft	412	801	999	1288	1997

Plotting this system head curve on the existing pump curve shows that the system head curve for the looped pipe needs to move to the right to obtain the 2800 gpm point of intersection with the pump head curve. We therefore need to increase the looped length further. By trial and error, the correct loop length required is 42 miles, and the corresponding system head curve is as follows:

Q gpm	1000	2100	2500	3000	4000
H_s ft	357	584	700	869	1283

Plotting this system head curve on the existing pump curve shows that the operating point is at the desired flow rate of 2800 gpm, as seen in Figure 8.18. Therefore, in the looped option 42 mi out of 54 mi must be looped to obtain the desired flow rate of 4000 bbl/h (2800 gpm). Next, the capital and operating costs of the two options will be calculated.

At $1500/ton for pipe material and $50/ft for pipe installation, the 42-mi loop cost is as follows:

From Appendix E, NPS 16, 0.250 in. wall thickness pipe weighs 42.05 lb/ft
Pipe material cost = 1500 × 42 × 5280 × 42.05/2000 = $6,993,756
Pipe labor cost = 50 × 42 × 5280 = $11,088,000
Total capital cost of the looping option = 6,993,756 + 11,088,000 = $18,081,756

Next, we calculate the cost of a new pump option.

The new pump and motor are expected to cost $300,000. The motor HP required was calculated earlier as 1035 HP. At the initial flow rate of 3000 bbl/h, the motor HP was calculated in Example 8.8 as 462/0.95 = 487 HP. Therefore, the incremental power cost for the new pump option is

$$= (1035 - 487) \times 0.746 \times 24 \times 365 \times 0.1 = \$358,116$$

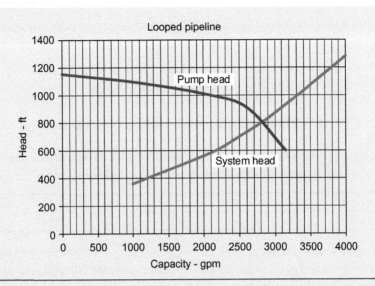

Figure 8.18 Looped pipeline.

With the looping option, the original pump is used to pump at 4000 bbl/h (2800 gpm). At this operating point, the BHP required is

$$BHP = 2800 \times 819 \times 0.74/(3960 \times 0.75) = 572$$

Considering motor efficiency of 95%:

$$Motor\ HP = 572/0.95 = 603$$

Therefore, the incremental power cost for the looping option is

$$= (603 - 487) \times 0.746 \times 24 \times 365 \times 0.1 = \$75{,}806$$

The two options can be compared as follows:

	New Pump Option	Looping Option
Capital Cost	$300,000	$18,081,756
Annual Cost	$358,116	$75,806

It can be seen that the new pump option is more cost effective and thus is the better of the two options.

EXAMPLE 8.11 SI UNITS

The Sorrento Pipeline Company historically has been transporting light crude oil from Florence to Milan via a 50 km, DN 600 pipeline. Recently, the company decided to transport heavy crude oil with a specific gravity of 0.895 and viscosity of 450 cSt at 20°C. The heavy crude will be received at the Florence tanks at 20°C and will be pumped out using the existing 2700 kW electric motor–driven centrifugal pump with the following water performance data:

Q m³/h	0	500	1000	1500	1800
H m	680	650	560	395	200
E%	0	52	71	80	78

Compared to the light crude, this pump will only be able to pump the heavy crude at a rate of 1080 m³/h. Due to the higher viscosity of the heavy crude, the pump performance is degraded and the efficiency at the 1080 m³/h flow rate is only 54%. The system head curve for the heavy crude based on the inlet temperature of 20°C is as follows:

Q m³/h	500	1000	1200	1500	1800
H_s m	372	486	532	601	777

In order to increase the pipeline flow rate, it is proposed to heat the crude oil to 35°C at Florence before pumping, thereby reducing the viscosity and hence the pressure drop in the pipeline. The crude oil heater will be purchased and installed upstream of the pump. The heated product entering the pump is at a temperature of 35°C, has a specific gravity of 0.825, and has a viscosity of 9.8 cSt at this temperature. The specific heat of the crude oil at 35°C is 1.88 kJ/kg/°C. The system head curve for the heated heavy crude based on the pump inlet temperature of 35°C is as follows:

Q m³/h	500	1000	1200	1500	1800
H_s m	291	361	399	467	547

It is estimated that the heater will cost $250,000, including material and labor costs. The annual maintenance of the heater will be $60,000. The natural gas used for heating the crude oil is expected to cost $5/TJ. (One Tera-Joule (TJ) = 1000 MJoule (MJ)). The heater efficiency is 80%. The incremental volume pumped through the pipeline from Florence to Milan will result in an additional revenue of $1.20/m³. Determine the economics associated with this project and the ROR anticipated for a 10-year project life.

Solution

Here are the steps for solving this problem:

1. Correct the pump performance for the high-viscosity crude oil at 20°C, using the Hydraulic Institute chart, as described in Chapter 3.

2. Plot the system head curve at 20°C, along with the corrected pump head curve to obtain the operating point (Q_1, H_1, E_1) for the 20°C pumping temperature. This will establish the flow rate and power required without crude oil heating.

3. Next, assume the heater is installed that will raise the crude oil temperature at the pump inlet from 20°C to 35°C. Since the viscosity at 35°C is less than 10 cSt, the pump curve does not require correction. The water performance curve given can be used.

4. Plot the given system head curve at 35°C and determine the new operating point (Q_2, H_2, E_2) on the water performance curves.

5. From the values of Q_1 and Q_2 and the corresponding pump heads and efficiencies, calculate the power required for the unheated and heated scenarios.

6. Calculate the heater duty in kJ/h based on the flow rate of Q_2, the liquid specific heat, and the temperature rise of $(35-20)$°C.

7. Calculate the total annual operating cost for the heater, including gas cost and heater maintenance cost.

8. Calculate the incremental annual power cost to run the pump at the higher flow rate Q_2.

9. Calculate the incremental annual revenue at the higher flow rate.

10. Tabulate the capital cost and the annual operating costs and determine the ROR for the project.

The pump curve is corrected for high viscosity as explained in Chapter 3. The system head curve for the 450 cSt crude at 20°C is plotted along with the corrected pump curve as shown in Figure 8.19. The operating point is as follows:

$$Q_1 = 1080 \text{ m}^3/\text{h} \quad H_1 = 514 \text{ m and } E_1 = 54.32\%$$

Next, the system head curve for the heated crude (9.8 cSt at 35°C) is plotted along with the uncorrected pump curve as shown in Figure 8.20. The operating point is

$$Q_2 = 1390 \text{ m}^3/\text{h} \quad H_2 = 447 \text{ m and } E_2 = 79.13\%$$

The power required at $Q_1 = 1080 \text{ m}^3/\text{h}$ is calculated using Equation (2.6) as

Power $= 1080 \times 514 \times 0.895/(367.46 \times 0.5432) = 2489 \text{ kW}$

Similarly, at the higher flow rate of $Q_2 = 1390 \text{ m}^3/\text{h}$

Power $= 1390 \times 447 \times 0.825/(367.46 \times 0.7913) = 1763 \text{ kW}$

Applying a motor efficiency of 95%, the incremental motor power required is

Incremental motor power $= (1763 - 2489)/0.95 = -764 \text{ kW}$

Thus, in the heated crude scenario, at the higher flow rate of 1390 m³/h, less motor power is required due to the lower crude oil viscosity and the uncorrected pump performance.

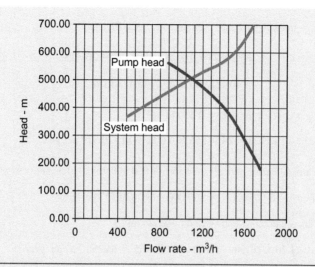

Figure 8.19 System curve for 450 cSt crude and corrected pump head curve.

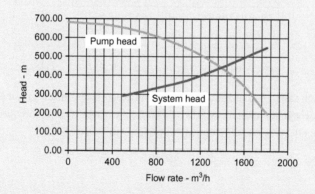

Figure 8.20 System curve for 9.8 cSt crude and uncorrected pump head curve.

Annual savings in electric power at $0.10 per kWh is

$$764 \times 24 \times 365 \times 0.10 = \$669,264 \text{ per year}$$

The heater duty is calculated next.

Crude oil mass flow rate at $1390 \text{m}^3/\text{h} = 1390 \times (1000 \times 0.825) = 1.14675 \times 10^6 \text{kg/h}$ based on the crude specific gravity of 0.825 and water density of 1000kg/m^3.

Heater duty = mass $\times$ sp. heat $\times$ temperature rise/efficiency

$$= 1.14675 \times 10^6 \times 1.88 \times (35 - 20)/0.8$$

$$= 40{,}422{,}938 \text{ kJ/h} = 40{,}423 \text{ MJ/h}$$

The annual gas cost for the heater at $5 /TJ is

$$40{,}423 \times 24 \times 365 \times 0.005 = \$1{,}770{,}527 \text{ per year}$$

Adding the annual heater maintenance cost of $60,000, the total heater cost per year is

Total heater cost = $1,770,527 + $60,000 = $1,830,527 per year

The annual incremental revenue based on $1.20 /m^3 is

$$1.20 \times (1390 - 1080) \times 24 \times 365 = \$3{,}258{,}720 \text{ per year}$$

The analysis can be summarized as follows:

	Heated Crude
Capital Cost	$250,000
Annual Cost	$1,830,527 ($669,264)
Annual Revenue	$3,258,720
Net Annual Revenue	$2,097,457

It can be seen that the installation of the crude oil heater results in a net revenue of over $2 million. The initial capital cost of $250,000 for the heater is easily recouped in less than two months. The rate of return on this project will be over 800%.

Most examples and practice problems in each chapter can be easily simulated using a simulation program that is available for download from the Elsevier website. The answers to the practice problems in each chapter are also available for review and can be downloaded from the same website. Although the simulation program is adequate for solving most problems, there are affordable, powerful software programs that can be used for centrifugal pump simulation. In the next chapter, we will review and discuss the simulation of centrifugal pumps using software developed by the author, titled PUMPCALC, that is available from SYSTEK Technologies, Inc. (*www.systek.us*).

Summary

In this chapter we reviewed several applications of centrifugal pumps in series and parallel configurations. We discussed how to generate the combined H-Q and E-Q curves for identical and unequal pumps in series and parallel. We looked at combined pump curves in conjunction with system head curves, and the impact of shutting down one pump. Several case studies involving alternatives for increasing pipeline capacity by changing pump impellers, adding pumps, and looping an existing pipeline were reviewed. In each case, we discussed the capital cost and annual operating cost and selected the option that was the most cost-effective. The rate of return on capital employed based on the project life was also discussed in some examples.

Problems

8.1 Three identical pumps each with H-Q data as shown are installed in series. Determine the combined pump head curve.

Q L/s	0	18	36	45	54
H m	102	97	81	70	60

8.2 Two unequal pumps with the following H-Q characteristics are installed in parallel configuration.

Pump A

Q gpm	1200	2160	2520	2880
H ft	1900	1550	1390	1180
E%	62	82	81	77

Pump B

Q gpm	1200	2160	2520	2880
H ft	2300	2200	1900	1400
E%	61	79	80	76

Determine the combined pump head and efficiency curves.

8.3 The system head curve for a pipeline using the three pumps in series in Problem 8.1 is defined by the equation

$$H_s = 59.6 + 2.3Q^2,$$

where Q is the capacity in L/s.

What is the operating point (Q, H) for this system? If one pump shuts down, what flow rate can be obtained with the remaining two pumps?

8.4 Using the two unequal parallel pumps in Problem 8.2, determine the operating point for the system head curve defined by the equation

$$H_s = 259 + 10.6Q^2,$$

where Q is the capacity in gpm.

If one pump shuts down, what flow rate can be obtained with the remaining pump?

8.5 A water pipeline is 120 km long. It is constructed of 1200 mm inside diameter cement pipe, with a Hazen-Williams C factor = 130. The present flow rate is 7400 m³/hr. To increase the flow rate to 10,500 m³/hr, two options are considered. One option is to install a second pump in parallel with the existing unit. The second option is to locate and install an intermediate booster pump. Determine the economics associated with these options and calculate the ROR for a 10-year project life. Assume a new pump station will cost $2000 per kW installed.

8.6 An existing pump station on a crude oil pipeline uses a turbine-driven centrifugal pump. Due to stricter environmental regulations and increased maintenance and operating costs, the company proposes to replace the turbine drive with an electric motor. Two options are to be investigated. The first option is to install a constant speed electric motor drive. The second option is to install a VFD pump. Discuss the approach to be followed in evaluating the two alternatives.

Chapter 9

Pump Simulation Using PUMPCALC Software

In this chapter we will demonstrate the simulation of various centrifugal pump–related problems using the commercial software PUMPCALC (*www.systek.us*). Interested readers may visit the website for downloading an evaluation version of the software.

PUMPCALC is a centrifugal pump analysis program. It can be used to predict the performance of a centrifugal pump at various impeller sizes and speeds. It can be used to determine the impeller trim or pump speed necessary to achieve a particular design point (Q, H) using the Affinity Laws. Multiple pump performance in series and parallel configurations can be modeled. For high-viscosity liquids, the viscosity corrected pump performance can be generated from the water performance curves using the

DOI: 10.1016/B978-1-85617-828-0.00009-3

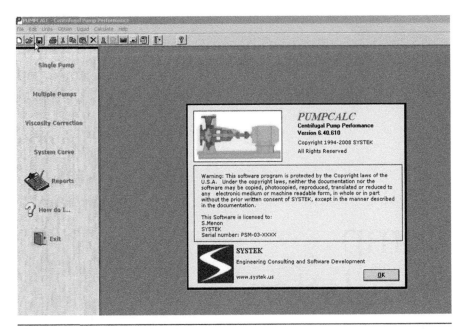

Figure 9.1 PUMPCALC main screen.

Hydraulic Institute method. Pump head curve versus system head curve and the pump operating point may be simulated as well. In the next few pages, the main features of PUMPCALC will be illustrated by simulating some examples taken from previous chapters. In addition to the features described in this chapter, PUMPCALC can be used to simulate system head curves, in conjunction with the pump head curves.

When you first launch the PUMPCALC program, the copyright screen appears as shown in Figure 9.1. The main screen shows the different simulation options on the left vertical panel, such as Single Pump, Multiple Pumps, Viscosity Correction, and System Curve. We will first choose the Single Pump option to simulate an example of a single pump similar to Example 2.6.

Single Pump Simulation

We will first select File | New from the upper menu bar to input the pump curve data given. A blank screen for entering the capacity, head, and efficiency data will be displayed. Enter the given pump curve data as shown in Figure 9.2. After entering the data, save the file as PUMP26. The file name extension .PMP is automatically added by the program.

Next, click the Interpolate button for calculating the BHP values for the various flow rates. This will display the screen shown in Figure 9.3 for calculating the BHP.

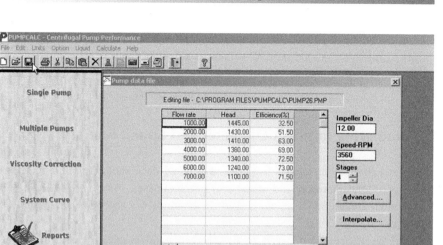

Figure 9.2 New pump data entry for Example 2.6.

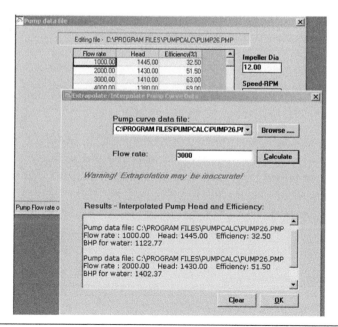

Figure 9.3 Calculating BHP for different capacities.

Enter the flow rates, 1000, 2000, and so on, and the BHP calculated is displayed as shown. Using this method, the BHP values are calculated and the table of BHP versus capacity can be obtained.

Simulating Impeller Diameter and Speed Change

In the next simulation, we will use the Example 6.1 data where the pump performance at different diameter and speed will be determined. As before, using File | New option, create the pump curve data for Example 6.1 and name the pump file PUMP61.PMP (see Figure 9.4).

Since the 12-inch impeller is to be changed to a 13-inch impeller, click the *Multiple curves...* button that will display the screen shown in Figure 9.5. Choose the different diameters option, and set the pump speed to 3560 RPM. Enter the two diameters 12 in. and 13 in. and click the OK button. The simulation will be completed, and the head curves for the two diameters will be plotted as in Figure 9.6.

Next, we will simulate the speed change, keeping the impeller size fixed at 12 inches. In this case, choose the different speeds option and enter the two speeds as shown in Figure 9.7.

Enter the two speeds 3560 and 4000 RPM, set the diameter at 12 inches, and click the OK button. The simulation will be completed, and the head curves for the two speeds will be plotted as in Figure 9.8.

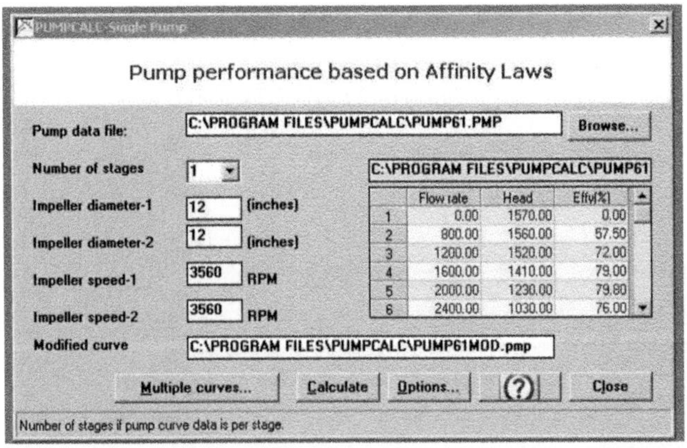

Figure 9.4 Performance based on Affinity Laws.

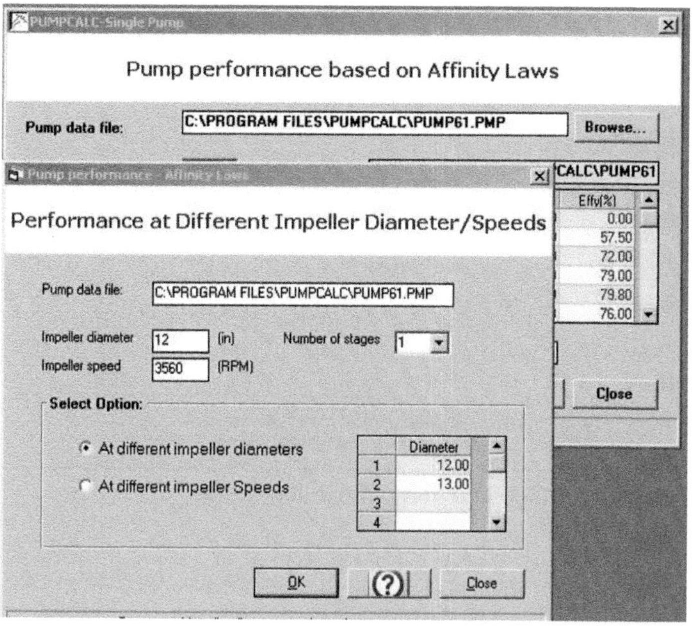

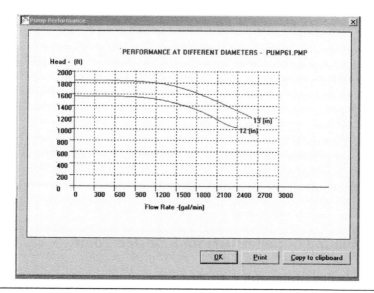

Figure 9.5   Performance at different impeller size or speed.

Figure 9.6   Head curves for two impeller sizes.

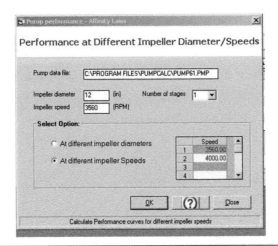

Figure 9.7 Performance at different impeller speed.

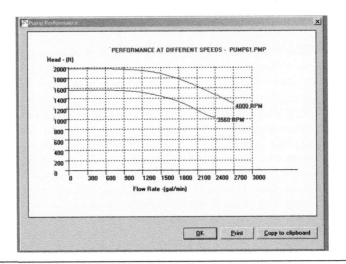

Figure 9.8 Head curves for two speeds.

Simulating Impeller Trim for a Design Point

We will use the Example 6.3 data for calculating the impeller size required to achieve a desired operating point of Q = 1900, H = 1680. As before, the pump curve data are entered and saved as a file named PUMP63.PMP. Next click the *Options* button to display the screen in Figure 9.9 for the Design point option.

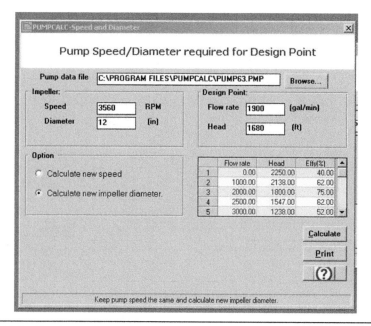

Figure 9.9   Design point option.

Choose the *Calculate new impeller diameter* option and enter the design point desired: Q = 1900, H = 1680. Clicking *Calculate* will start simulation, and the result is displayed as in Figure 9.10.

Next, we will use the Example 6.4 data for calculating the impeller speed required to achieve a desired operating point Q = 450 m³/h, H = 300 m. This is in SI units, so we must first choose the proper units from the Units menu of the main screen. Choose the appropriate units for the head and flow rate. Next, as before, the pump curve data is entered and saved as a file named PUMP64.PMP. Next click the *Options* button to display the screen for the Design point option, as in Figure 9.11.

Choose the *Calculate new speed* option and enter the design point desired: Q = 450 m³/h, H = 300 m. Clicking the *Calculate* button will start simulation, and the result is displayed as in Figure 9.12.

It can be seen from these two simulations that the results are very close to what we obtained in Chapter 6. The results from PUMPCALC are more accurate, since it employs a more rigorous calculation methodology.

Viscosity Correction Example

Next, we will use the *Viscosity Correction* option to predict the performance of a pump with a high-viscosity liquid. Let us choose the same pump used in Example 6.3, named

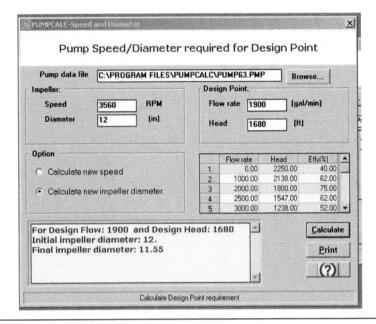

Figure 9.10 Impeller trim required for design point.

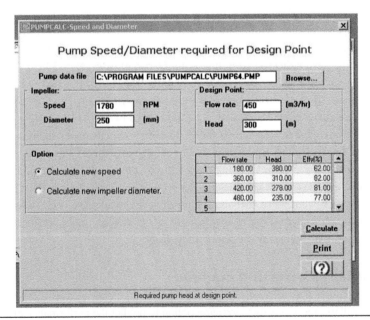

Figure 9.11 Design point option for speed change.

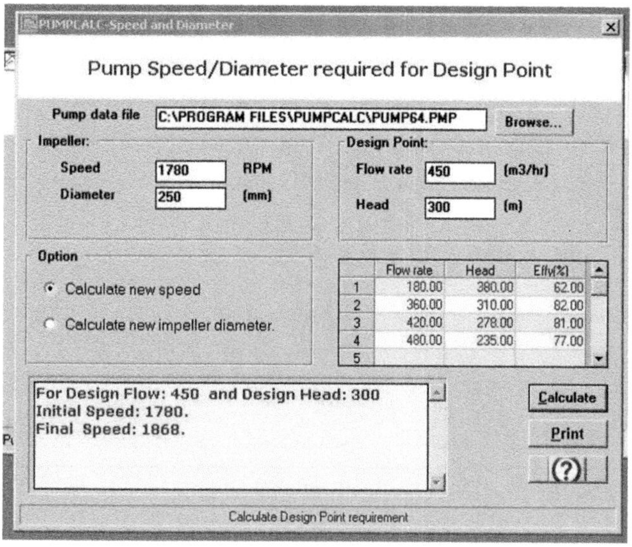

Figure 9.12 Impeller speed required for design point.

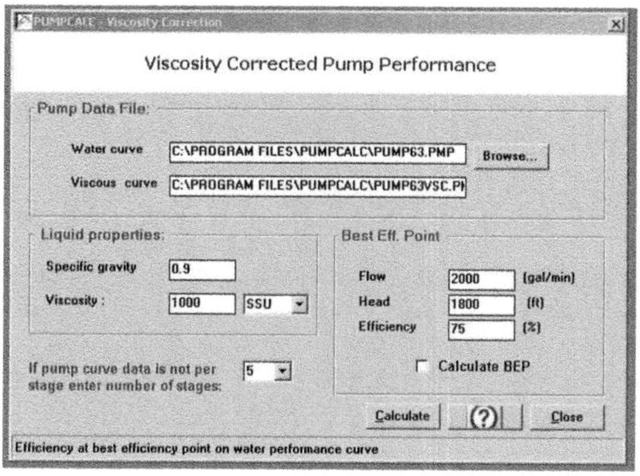

Figure 9.13 Viscosity correction data.

PUMP63.PMP, and determine the viscosity corrected performance when pumping a liquid with these properties: specific gravity = 0.9 and viscosity = 1000 SSU.

Select the *Viscosity Correction* option from the left panel and after browsing for and choosing the pump file PUMP63.PMP, the screen in Figure 9.13 is displayed. The viscous performance curve will have the name PUMP63VSC.PMP, which is automatically assigned when you tab over to the viscous curve data entry field, as shown. Enter the liquid properties and the number of stages of the pump, and check the box for *Calculate BEP.* Click the *Calculate* button, and the simulation starts.

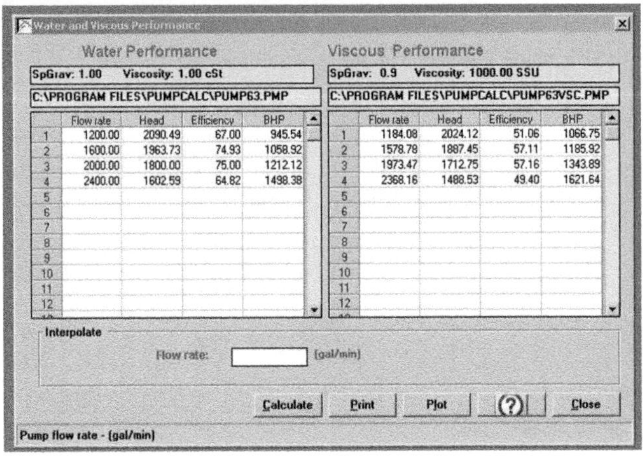

Water and Viscous Performance

Water Performance
SpGrav: 1.00   Viscosity: 1.00 cSt
C:\PROGRAM FILES\PUMPCALC\PUMP63.PMP

	Flow rate	Head	Efficiency	BHP
1	1200.00	2090.49	67.00	945.54
2	1600.00	1963.73	74.93	1058.92
3	2000.00	1800.00	75.00	1212.12
4	2400.00	1602.59	64.82	1498.38
5				
6				
7				
8				
9				
10				
11				
12				

Viscous Performance
SpGrav: 0.9   Viscosity: 1000.00 SSU
C:\PROGRAM FILES\PUMPCALC\PUMP63VSC.PMP

	Flow rate	Head	Efficiency	BHP
1	1184.08	2024.12	51.06	1066.75
2	1578.78	1897.45	57.11	1185.92
3	1973.47	1712.75	57.16	1343.89
4	2368.16	1488.53	49.40	1621.64
5				
6				
7				
8				
9				
10				
11				
12				

Interpolate
Flow rate: [   ] (gal/min)

Calculate   Print   Plot   (?)   Close

Pump flow rate - (gal/min)

Figure 9.14 Water and viscous performance data.

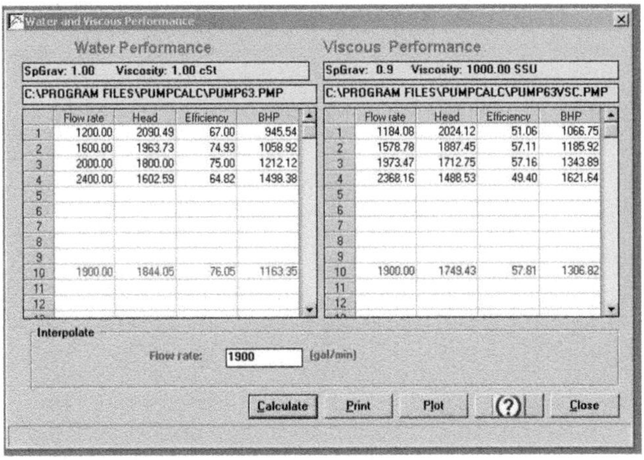

Water and Viscous Performance

Water Performance
SpGrav: 1.00   Viscosity: 1.00 cSt
C:\PROGRAM FILES\PUMPCALC\PUMP63.PMP

	Flow rate	Head	Efficiency	BHP
1	1200.00	2090.49	67.00	945.54
2	1600.00	1963.73	74.93	1058.92
3	2000.00	1800.00	75.00	1212.12
4	2400.00	1602.59	64.82	1498.38
5				
6				
7				
8				
9				
10	1900.00	1844.05	76.05	1163.35
11				
12				

Viscous Performance
SpGrav: 0.9   Viscosity: 1000.00 SSU
C:\PROGRAM FILES\PUMPCALC\PUMP63VSC.PMP

	Flow rate	Head	Efficiency	BHP
1	1184.08	2024.12	51.06	1066.75
2	1578.78	1897.45	57.11	1185.92
3	1973.47	1712.75	57.16	1343.89
4	2368.16	1488.53	49.40	1621.64
5				
6				
7				
8				
9				
10	1900.00	1749.43	57.81	1306.82
11				
12				

Interpolate
Flow rate: [1900] (gal/min)

Calculate   Print   Plot   (?)   Close

Figure 9.15 Interpolated performance data.

The results are displayed as in Figure. 9.14, where the water performance curve is shown on the left side and the viscosity corrected performance of the pump is on the right side. If we want to determine the head available for a particular capacity, the capacity value is entered in the text box below the performance data, and the program will then calculate both water performance as well as the viscous performance by interpolation.

Figure 9.15 shows the interpolated values for Q = 1900 gpm for both the water curve and the viscous curve. Click the Plot button, and the program plots the water performance curve and the viscous curve as in Figure. 9.16.

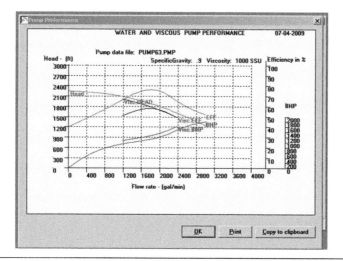

Figure 9.16   Water performance and viscous performance data.

Summary

In this chapter we introduced you to the PUMPCALC simulation software. We reviewed some examples of single-pump performance and performance at different speeds and impeller diameters. Also, we illustrated the use of PUMPCALC to determine the pump impeller trim or speed required to achieve a design point. An example of the viscosity corrected performance was also simulated. The reader is advised to download an evaluation copy of the software from the software publisher's website (*www.systek.us*) to explore the many features of PUMPCALC.

Appendix A

Summary of Formulas

Chapter 1

Conversion from specific gravity to API gravity and vice versa

$$\text{Specific gravity } Sg = 141.5/(131.5 + \text{API}) \tag{1.2a}$$

$$\text{API} = 141.5/Sg - 131.5 \tag{1.2b}$$

© 2010 E. Shashi Menon. Published by Elsevier Inc. All right reserved.
DOI: 10.1016/B978-1-85617-828-0.00012-3

Viscosity conversion from SSU and SSF to kinematic viscosity, cSt

$$\nu = 0.226 \times SSU - 195/SSU \text{ for } 32 \leq SSU \leq 100 \qquad (1.6)$$

$$\nu = 0.220 \times SSU - 135/SSU \text{ for } SSU > 100 \qquad (1.7)$$

$$\nu = 2.24 \times SSF - 184/SSF \text{ for } 25 \leq SSF \leq 40 \qquad (1.8)$$

$$\nu = 2.16 \times SSF - 60/SSF \text{ for } SSF > 40 \qquad (1.9)$$

where ν is the viscosity in centistokes at a particular temperature.

Pressure P at a depth h below the free surface is as follows.

USCS Units

$$P = h \times Sg/2.31 \qquad (1.11)$$

where

P – pressure, psig
h – depth below free surface of liquid, ft
Sg – specific gravity of liquid, dimensionless

SI Units

$$P = h \times Sg/0.102 \qquad (1.12)$$

where

P – pressure, kPa,
Sg – specific gravity of liquid, dimensionless
h – depth below free surface of liquid, m

Bernoulli's Equation

$$P_A/\gamma + V_A^2/2g + Z_A - h_f + h_p = P_B/\gamma + V_B^2/2g + Z_B \qquad (1.15)$$

where, P_A, V_A, and Z_A are the pressure, velocity, and elevation head at point A, respectively; subscript B refers to the point B; γ is the specific weight of liquid; h_f is the frictional pressure drop; and h_p is the pump head.

Chapter 2

The H-Q curve can be represented by the following equation:

$$H = a_0 + a_1Q + a_2Q^2 \qquad (2.1)$$

where a_0, a_1, and a_2 are constants.

The E versus Q curve can also be represented by a parabolic equation as follows:

$$E = b_0 + b_1Q + b_2Q^2 \qquad (2.2)$$

where b_0, b_1, and b_2 are constants for the pump.

In the USCS units, the HHP is calculated as follows:

$$HHP = Q \times H \times Sg/3960 \qquad (2.3)$$

where

Q – capacity, gal/min
H – head, ft
Sg – specific gravity of liquid pumped, dimensionless

In SI units, the hydraulic power required in kW is as follows:

$$\text{Hydraulic power (kW)} = Q \times H \times Sg/(367.46) \qquad (2.4)$$

where

Q – capacity, m^3/h
H – head, m
Sg – specific gravity of liquid pumped, dimensionless

In USCS units, BHP Calculation

$$BHP = Q \times H \times Sg/(3960E) \qquad (2.5)$$

where

Q – capacity, gal/min
H – head, ft
Sg – specific gravity of liquid pumped, dimensionless
E – pump efficiency (decimal value, less than 1.0)

In SI units, the pump power required in kW is as follows:

$$\text{Power} = Q \times H \times Sg/(367.46E) \qquad (2.6)$$

where

Q – capacity, m^3/h

H – head, m

Sg – specific gravity of liquid pumped, dimensionless

E – pump efficiency (decimal value, less than 1.0)

The synchronous speed of an electric motor can be calculated as follows:

$$N_S = 120 \times f/p \qquad (2.9)$$

where

f – electrical frequency

p – number of poles

The specific speed is calculated using the formula:

$$N_S = N\,Q^{\frac{1}{2}}/H^{\frac{3}{4}} \qquad (2.10)$$

where

N_S – specific speed of the pump

N – impeller speed, RPM

Q – capacity at BEP, gal/min

H – head at BEP, ft

In SI units, the same Equation (2.10) is used for specific speed, except Q will be in m^3/h and H in m.

Suction specific speed is calculated as follows:

$$N_{SS} = N\,Q^{\frac{1}{2}}/(NPSH_R)^{\frac{3}{4}} \qquad (2.11)$$

where

N_{SS} – suction specific speed of the pump

N – impeller speed, RPM

Q – capacity at BEP, gal/min

$NPSH_R$ – NPSH required at BEP, ft

In SI units, the same formula is used, with Q in m^3/h and $NPSH_R$ in m.

Chapter 3

Temperature rise calculation:

USCS UNITS

$$\Delta T = H\,(1/E - 1)/(778Cp) \qquad (3.1)$$

where

ΔT – temperature rise of liquid from suction to discharge of pump

H – head at the operating point, ft
E – efficiency at the operating point, (decimal value, less than 1.0)
Cp – liquid specific heat, Btu/lb/°F

SI Units

$$\Delta T = H \, (1/E - 1)/(101.94 Cp) \tag{3.2}$$

where
ΔT – temperature rise of liquid from suction to discharge of pump, °C
H – head at the operating point, m
E – efficiency at the operating point (decimal value, less than 1.0)
Cp – liquid specific heat, kJ/kg/°C

The temperature rise of the liquid per minute when the pump is operated with a closed valve.

USCS Units

$$\Delta T/min = 42.42 \, BHP_0/(MCp) \tag{3.3}$$

where
ΔT – temperature rise, °F per min
BHP_0 – BHP required under shutoff conditions
M – amount of liquid contained in pump, lb
Cp – liquid specific heat, Btu/lb/°F

In the SI units:

$$\Delta T/min = 59.98 \, P_0/(MCp) \tag{3.4}$$

where
ΔT – temperature rise, °C per min
P_0 – power required under shut off conditions, kW
M – amount of liquid contained in pump, kg
Cp – liquid specific heat, kJ/kg/°C.

Chapter 4

Velocity of flow:

USCS units

$$V = 0.4085 \, Q/D^2 \tag{4.4}$$

where
  V – average flow velocity, ft/s
  Q – flow rate, gal/min
  D – inside diameter of pipe, inches

$$V = 0.2859 \ Q/D^2 \tag{4.5}$$

where
  V – average flow velocity, ft/s
  Q – flow rate, bbl/h
  D – inside diameter of pipe, inches

SI Units

$$V = 353.6777 \ Q/D^2 \tag{4.6}$$

where
  V – average flow velocity, m/s
  Q – flow rate, m^3/h
  D – inside diameter of pipe, mm

$$V = 1273.242 \ Q/D^2 \tag{4.6a}$$

where
  V – average flow velocity, m/s
  Q – flow rate, L/s
  D – inside diameter of pipe, mm

Reynolds number calculation:

USCS Units

$$R = 3160 \ Q/(\nu D) \tag{4.8}$$

where
  R – Reynolds number, dimensionless
  Q – flow rate, gal/min
  ν – kinematic viscosity of the liquid, cSt
  D – inside diameter of pipe, inch

$$R = 2214 \ Q/(\nu D) \tag{4.9}$$

where

R – Reynolds number, dimensionless
Q – flow rate, bbl/h
ν – kinematic viscosity of the liquid, cSt
D – inside diameter of pipe, inch

SI UNITS

$$R = 353,678 \; Q/(\nu D) \tag{4.10}$$

where

R – Reynolds number, dimensionless
Q – flow rate, m³/h
ν – kinematic viscosity of the liquid, cSt
D – inside diameter of pipe, mm

$$R = 1.2732 \times 10^6 \; Q/(\nu D) \tag{4.10a}$$

where

R – Reynolds number, dimensionless
Q – flow rate, L/s
ν – kinematic viscosity of the liquid, cSt
D – inside diameter of pipe, mm

Darcy or Darcy-Weisbach equation

USCS UNITS

$$h = f \, (L/D) \, V^2/2g \tag{4.11}$$

where

h – head loss due to friction, ft
f – friction factor, dimensionless
L – pipe length, ft
D – inside diameter of pipe, ft
V – average flow velocity, ft/s
g – acceleration due to gravity = 32.2 ft/s²

SI UNITS

$$h = f \, (L/D) \, V^2/2g \tag{4.11a}$$

where

 h – head loss due to friction, m
 f – friction factor, dimensionless
 L – pipe length, m
 D – inside diameter of pipe, m
 V – average flow velocity, m/s
 g – acceleration due to gravity = 9.81 m/s^2

Pressure drop:

USCS Units

$$P_m = 71.1475 \, fQ^2Sg/D^5 \tag{4.12}$$

$$h_m = 164.351 \, fQ^2/D^5 \tag{4.12a}$$

where

 P_m – pressure drop, psi/mi
 h_m – head loss, ft of liquid/mi
 f – friction factor, dimensionless
 Q – flow rate, gal/min
 Sg – specific gravity of liquid, dimensionless
 D – inside diameter of pipe, inch

When flow rate is in bbl/h, the equation becomes

$$P_m = 34.8625 \, fQ^2Sg/D^5 \tag{4.13}$$

$$h_m = 80.532 \, fQ^2/D^5 \tag{4.13a}$$

where

 P_m – pressure drop, psi/mi
 h_m – head loss, ft of liquid/mi
 f – friction factor, dimensionless
 Q – flow rate, bbl/h
 Sg – specific gravity of liquid, dimensionless
 D – inside diameter of pipe, inch

SI Units

$$P_{km} = 6.2475 \times 10^{10} \, f \, Q^2 \, (Sg/D^5) \tag{4.14}$$

$$h_{km} = 6.372 \times 10^9 \, f \, Q^2 \, (1/D^5) \qquad (4.14a)$$

where

P_{km} – pressure drop, kPa/km

h_{km} – head loss, m of liquid/km

f – friction factor, dimensionless

Q – flow rate, m^3/h

Sg – specific gravity of liquid, dimensionless

D – inside diameter of pipe, mm

Pressure drop using transmission factor F:

USCS Units

$$P_m = 284.59 \, (Q/F)^2 \, Sg/D^5 \qquad \text{for Q in gal/min} \qquad (4.17)$$

$$P_m = 139.45 \, (Q/F)^2 \, Sg/D^5 \qquad \text{for Q in bbl/h} \qquad (4.18)$$

SI Units

$$P_{km} = 24.99 \times 10^{10} \, (Q/F)^2 (Sg/D^5) \qquad \text{for Q in } m^3/hr \qquad (4.19)$$

Colebrook-White equation for friction factor:

$$1/\sqrt{f} = -2 \, Log_{10}[(e/3.7D) + 2.51/(R\sqrt{f})] \qquad (4.21)$$

where

f – friction factor, dimensionless

D – inside diameter of pipe, inch

e – absolute roughness of pipe, inch

R – Reynolds number, dimensionless

In SI units the same equation can be used if D and e are both in mm.

P. K. Swamee and A. K. Jain proposed an explicit equation for the friction factor in 1976 in the *Journal of the Hydraulics Division of ASCE*.

$$f = 0.25/[Log_{10}(e/3.7D + 5.74/R^{0.9})]^2 \qquad (4.22)$$

Another explicit equation for the friction factor, proposed by Stuart Churchill, was reported in *Chemical Engineering* in November 1977. It requires the calculation of parameters A and B, which are functions of the Reynolds number R, as follows:

$$f = [(8/R)^{12} + 1/(A + B)^{3/2}]^{1/12} \qquad (4.23)$$

where

$$A = [2.457\text{Log}_e(1/((7/R)^{0.9} + (0.27e/D))]^{16} \qquad (4.24)$$

$$B = (37530/R)^{16} \qquad (4.25)$$

Hazen-Williams equation:

USCS Units

$$Q = 6.7547 \times 10^{-3}(C)(D)^{2.63}(h)^{0.54} \qquad (4.27)$$

where
- Q – flow rate, gal/min
- C – Hazen-Williams C factor
- D – inside diameter of pipe, inch
- h – head loss due to friction per 1000 ft of pipe, ft

Equation (4.27) can be transformed to solve for the head loss h in terms of flow rate Q and other variables as

$$h = 1.0461 \times 10^4 (Q/C)^{1.852}(1/D)^{4.87} \qquad (4.27a)$$

SI Units

$$Q = 9.0379 \times 10^{-8}(C)(D)^{2.63}(P_{km}/Sg)^{0.54} \qquad (4.28)$$

$$h_{km} = 1.1323 \times 10^{12}(Q/C)^{1.852}(1/D)^{4.87} \qquad (4.28a)$$

where
- P_{km} – pressure drop due to friction, kPa/km
- h_{km} – pressure drop due to friction, m/km

Q – flow rate, m³/h
C – Hazen-Williams C factor
D – pipe inside diameter, mm
Sg – specific gravity of liquid, dimensionless

USCS AND SI UNITS

$$\text{Velocity head} = V^2/2g \qquad (4.29)$$

where
V – velocity of flow, ft or m
g – acceleration due to gravity, ft/s² or m/s²

Friction loss in a valve or fitting:

$$h_f = K(V^2/2g) \qquad (4.30)$$

The head loss due to sudden enlargement:

$$K = [1 - (D_1/D_2)^2]^2 \qquad (4.31)$$

The ratio A_2/A_1 is easily calculated from the diameter ratio D_2/D_1 as follows:

$$A_2/A_1 = (D_2/D_1)^2 \qquad (4.32)$$

Total equivalent length:

$$L_E = L_A + L_B(D_A/D_B)^5 \qquad (4.34)$$

Total equivalent diameter for parallel pipe:

$$D_E = D_1(Q/Q_1)^{0.4} \qquad (4.42a)$$

$$D_E = D_2(L_1/L_2)^{0.2}(Q/Q_2)^{0.4} \qquad (4.43a)$$

Chapter 6

Affinity Laws equation for pump impeller diameter change:

$$Q_2/Q_1 = D_2/D_1 \qquad (6.1)$$

$$H_2/H_1 = (D_2/D_1)^2 \qquad (6.2)$$

Affinity Laws for speed change:

$$Q_2/Q_1 = N_2/N_1 \qquad (6.3)$$

$$H_2/H_1 = (N_2/N_1)^2 \qquad (6.4)$$

Affinity Laws equation for the change in power required when impeller diameter or speed is changed:

For diameter change:

$$BHP_2/BHP_1 = (D_2/D_1)^3 \qquad (6.5)$$

For speed change:

$$BHP_2/BHP_1 = (N_2/N_1)^3 \qquad (6.6)$$

Correction for impeller trim:

$$y = (5/6)(x + 20) \qquad (6.7)$$

where
  x = calculated trim for impeller diameter, %
  y = corrected trim for impeller diameter, %

Chapter 7

$NPSH_A$ can be calculated using the equation

$$NPSH_A = h_a + h_s - h_f - h_{vp} \qquad (7.1)$$

where

NPSH$_A$ – available net positive suction head, ft of liquid
h$_a$ – atmospheric pressure at surface of liquid in tank, ft
h$_s$ – suction head from liquid level in tank to pump suction, ft
h$_f$ – frictional head loss in suction piping, ft
h$_{vp}$ – vapor pressure of liquid at pumping temperature, ft

Chapter 8

USCS UNITS

In combined performance, the efficiency values are calculated as follows:

$$Q_T H_T/E_T = Q_1 H_1/E_1 + Q_2 H_2/E_2 + Q_3 H_3/E_3 \qquad (8.1)$$

where subscript T is used for the total or combined performance, and 1, 2, and 3 are for each of the three pumps.

Simplifying Equation (8.1), by setting $Q_1 = Q_2 = Q_3 = Q_T$ for series pumps and solving for the combined efficiency E_T, we get

$$E_T = H_T/(H_1/E_1 + H_2/E_2 + H_3/E_3) \qquad (8.2)$$

SI UNITS

In combined performance, the total power is calculated using the following equation:

$$Q_T H_T/E_T = Q_1 H_1/E_1 + Q_2 H_2/E_2 \qquad (8.3)$$

where subscript T is used for the combined performance, and 1, 2, are for each of the two pumps.

Simplifying Equation (8.3) by setting $H_1 = H_2 = H_T$ for parallel pumps and solving for the combined efficiency E_T, we get

$$E_T = Q_T/(Q_1/E_1 + Q_2/E_2) \qquad (8.4)$$

Appendix B

Units and Conversion Factors

DOI: 10.1016/B978-1-85617-828-0.00013-5

SI Units—Systeme International Units (modified metric) USCS Units—U.S. Customary System of Units

Item	SI Units	USCS Units	SI to USCS Conversion	USCS to SI Conversion
Mass	kilogram (kg)	slug (slug)	1 kg = 0.0685 slug	1 lb = 0.45359 kg
	metric tonne (t) = 1000 kg	pound mass (lbm)	1 kg = 2.205 lb	1 slug = 14.594 kg
		1 U.S. ton = 2000 lb	1 t = 1.1023 US ton	1 U.S. ton = 0.9072 t
		1 long ton = 2240 lb	1 t = 0.9842 long ton	1 long ton = 1.016 t
Length	millimeter (mm)	inch (in)	1 mm = 0.0394 in	1 in. = 25.4 mm
	1 meter (m) = 1000 mm	1 foot (ft) = 12 in	1 m = 3.2808 ft	1 ft = 0.3048 m
	1 kilometer (km) = 1000 m	1 mile (mi) = 5,280 ft	1 km = 0.6214 mi	1 mi = 1.609 km
Area	square meter (m^2)	square foot (ft^2)	1 m^2 = 10.764 ft^2	1 ft^2 = 0.0929 m^2
	1 hectare = 10,000 m^2	1 acre = 43,560 ft^2	1 hectare = 2.4711 acre	1 acre = 0.4047 hectare
Volume	cubic millimeter (mm^3)	cubic inch (in^3)	1 mm^3 = 6.1 × 10^{-5} in^3	1 in^3 = 16387.0 mm^3
	1 liter (L) = 1000 cm^3 (cc)	cubic foot (ft^3)	1 m^3 = 35.3134 ft^3	1 ft^3 = 0.02832 m^3
	1 cubic meter (m^3) = 1000 L	1 U.S. gallon (gal) = 231 in^3	1 L = 0.2642 gal	1 gal = 3.785 L
		1 barrel (bbl) = 42 gal	1 m^3 = 6.2905 bbl	1 bbl = 158.97 L = 0.15897 m^3
		1 ft^3 = 7.4805 gal		
		1 bbl = 5.6146 ft^3		

Quantity	SI Units	US Units	Conversion	
Density	kilogram/cubic meter (kg/m³)	slug per cubic foot (slug/ft³)	1 kg/m³ = 0.0019 slug/ft³	1 slug/ft³ = 515.38 kg/m³
Specific Weight	Newton per cubic meter (N/m³)	pound per cubic foot (lb/ft³)	1 N/m³ = 0.0064 lb/ft³	1 lb/ft³ = 157.09 N/m³
Viscosity (Absolute or Dynamic)	1 poise (P) = 0.1 Pa-s 1 centipoise (cP) = 0.01 P 1 poise = 1 dyne-s/cm² 1 poise = 0.1 N-s/m²	lb/ft-s lb-s/ft²	1 cP = 6.7197×10^{-4} lb/ft-s 1 N-s/m² = 0.0209 lb-s/ft² 1 poise = 0.00209 lb-s/ft²	1 lb-s/ft² = 47.88 N-s/m² 1 lb-s/ft² = 478.8 Poise
Viscosity (Kinematic)	m²/s stoke (S), centistoke (cSt)	ft²/s SSU*, SSF*	1 m²/s = 10.7639 ft²/s 1 cSt = 1.076×10^{-5} ft²/s	1 ft²/s = 0.092903 m²/s
Flow Rate	liter/minute (L/min) cubic meter/hour (m³/h)	cubic foot/second (ft³/s) gallon/minute (gal/min) barrel/hour (bbl/h) barrel/day (bbl/day)	1 L/min = 0.2642 gal/min 1 m³/h = 6.2905 bbl/h	1 gal/min = 3.7854 L/min 1 bbl/h = 0.159 m³/h
Force	Newton (N) = kg-m/s²	pound (lb)	1 N = 0.2248 lb	1 lb = 4.4482 N

(Continued)

(Continued)

Item	SI Units	USCS Units	SI to USCS Conversion	USCS to SI Conversion
Pressure	Pascal (Pa) = N/m^2 1 kiloPascal (kPa) = 1000 Pa 1 megaPascal (MPa) = 1000 kPa 1 bar = 100 kPa kilogram/sq. centimeter (kg/cm^2)	pound/square inch, lb/in^2 (psi) $1\ lb/ft^2$ = 144 psi	1 kPa = 0.145 psi 1 bar = 14.5 psi $1\ kg/cm^2$ = 14.22 psi	1 psi = 6.895 kPa 1 psi = 0.069 bar 1 psi = 0.0703 kg/cm^2
Velocity	meter/second (m/s)	ft/second (ft/s) mile/hour (mi/h) = 1.4667 ft/s	1 m/s = 3.281 ft/s	1 ft/s = 0.3048 m/s
Work and Energy	joule (J) = N-m	foot-pound (ft-lb) British thermal unit (Btu) 1 Btu = 778 ft-lb	1 kJ = 0.9478 Btu	1 Btu = 1055.0 J

Power	joule/second (J/s)	ft-lb/min	$1\,W = 3.4121\,Btu/h$	$1\,Btu/h = 0.2931\,W$
	watt (W) = J/s	Btu/hour		
	1 kilowatt (kW) = 1000 W	horsepower (HP)	$1\,kW = 1.3405\,HP$	$1\,HP = 0.746\,kW$
		$1\,HP = 33,000\,ft\text{-}lb/min$		
Temperature	degrees Celsius (°C)	degree Fahrenheit (°F)	$1°C = (°F - 32)/1.8$	$1°F = 9/5°C + 32$
	1 degrees Kelvin (K) = °C +273	1 degree Rankin (°R) = °F + 460	$1\,K = °R/1.8$	$1°R = 1.8\,K$
Thermal Conductivity	W/m/°C	Btu/h/ft/°F	$1\,W/m/°C = 0.5778\,Btu/h/ft/°F$	$1\,Btu/h/ft/°F = 1.7307\,W/m/°C$
Heat Transfer Coefficient	W/m²/°C	Btu/h/ft²/°F	$1\,W/m^2/°C = 0.1761\,Btu/h/ft^2/°F$	$1\,Btu/h/ft^2/°F = 5.6781\,W/m^2/°C$
Specific Heat	kJ/kg/°C	Btu/lb/°F	$1\,kJ/kg/°C = 0.2388\,Btu/lb/°F$	$1\,Btu/lb/°F = 4.1869\,kJ/kg/°C$

Note: Kinematic viscosity in SSU and SSF are converted to viscosity in cSt using the following formulas:

Centistokes = 0.226 × SSU − 195/SSU for 32 ≤ SSU ≤ 100

Centistokes = 0.220 × SSU − 135/SSU for SSU > 100

Centistokes = 2.24 × SSF − 184/SSF for 25 ≤ SSF ≤ 40

Centistokes = 2.16 × SSF − 60/SSF for SSF > 40

Appendix C

DOI: 10.1016/B978-1-85617-828-0.00014-7

Properties of Water—USCS Units

Temperature °F	Density lb/ft^3	Dynamic Viscosity lb-s/ft$^2 \times 10^6$	Kinematic Viscosity ft^2/s $\times 10^6$
32	62.4	36.6	18.9
40	62.4	32.3	16.7
50	62.4	27.2	14.0
60	62.4	23.5	12.1
70	62.3	20.4	10.5
80	62.2	17.7	9.15
90	62.1	16.0	8.29
100	62.0	14.2	7.37
110	61.9	12.6	6.55
120	61.7	11.4	5.94
130	61.5	10.5	5.49
140	61.4	9.6	5.03
150	61.2	8.9	4.68
160	61.0	8.3	4.38
170	60.8	7.7	4.07
180	60.6	7.23	3.84
190	60.4	6.80	3.62
200	60.1	6.25	3.35
212	59.8	5.89	3.17

Conversions

$1 \, \text{lb-s/ft}^2 = 47.88 \, \text{N-s/m}^2 = 478.8 \, \text{poise} = 4.788 \times 10^4 \, \text{cP}$

$1 \, \text{ft}^2/\text{s} = 929 \, \text{stoke} = 9.29 \times 10^4 \, \text{cSt}$

Properties of Water—SI Units

Temperature °C	Density kg/m^3	Dynamic Viscosity N-s/m$^2 \times 10^4$	Kinematic Viscosity m^2/s $\times 10^7$
0	1000	17.5	17.5
10	1000	13.0	13.0
20	998	10.2	10.2
30	996	8.0	8.03
40	992	6.51	6.56
50	988	5.41	5.48
60	984	4.60	4.67
70	978	4.02	4.11
80	971	3.50	3.60
90	965	3.11	3.22
100	958	2.82	2.94

Conversions

$1\,\text{N-s/m}^2 = 10\,\text{poise} = 1000\,\text{cP}$

$1\,\text{m}^2/\text{s} = 1 \times 10^4\,\text{stoke} = 1 \times 10^6\,\text{cSt}$

Appendix D

DOI: 10.1016/B978-1-85617-828-0.00015-9

Properties of Common Liquids

Products	API Gravity	Specific Gravity @ 60°F	Viscosity cSt @ 60°F	Reid Vapor Pressure psi
Regular Gasoline				
Summer Grade	62.0	0.7313	0.70	9.5
Interseasonal Grade	63.0	0.7275	0.70	11.5
Winter Grade	65.0	0.7201	0.70	13.5
Premium Gasoline				
Summer Grade	57.0	0.7467	0.70	9.5
Interseasonal Grade	58.0	0.7165	0.70	11.5
Winter Grade	66.0	0.7711	0.70	13.5
No. 1 Fuel Oil	42.0	0.8155	2.57	
No. 2 Fuel Oil	37.0	0.8392	3.90	
Kerosene	50.0	0.7796	2.17	
JP-4	52.0	0.7711	1.40	2.7
JP-5	44.5	0.8040	2.17	

Appendix E

DOI: 10.1016/B978-1-85617-828-0.00016-0

Properties of Circular Pipes—USCS Units

Nominal Pipe Size NPS	Outside Diameter in.	Schedule a	Schedule b	Schedule c	Wall Thickness in.	Inside Diameter in.	Flow Area in.²	Volume gal/ft	Pipe Weight lb/ft	Water Weight* lb/ft
½	0.84			5S	0.065	0.710	0.3957	0.02	0.54	0.17
	0.84			10S	0.083	0.674	0.3566	0.02	0.67	0.15
	0.84	40	Std	40S	0.109	0.622	0.3037	0.02	0.85	0.13
	0.84	80	XS	80S	0.147	0.546	0.2340	0.01	1.09	0.10
	0.84	160			0.187	0.466	0.1705	0.01	1.30	0.07
	0.84		XXS		0.294	0.252	0.0499	0.00	1.71	0.02
¾	1.05			5S	0.065	0.920	0.6644	0.03	0.68	0.29
	1.05			10S	0.083	0.884	0.6134	0.03	0.86	0.27
	1.05	40	Std	40S	0.113	0.824	0.5330	0.03	1.13	0.23
	1.05	80	XS	80S	0.154	0.742	0.4322	0.02	1.47	0.19
	1.05	160			0.218	0.614	0.2959	0.02	1.94	0.13
	1.05		XXS		0.308	0.434	0.1479	0.01	2.44	0.06
1	1.315			5S	0.065	1.185	1.1023	0.06	0.87	0.48
	1.315			10S	0.109	1.097	0.9447	0.05	1.40	0.41
	1.315	40	Std	40S	0.330	0.655	0.3368	0.02	3.47	0.15

	1.315	80	XS	80S	0.179	0.957	0.7189	0.04	2.17	0.31
	1.315	160			0.250	0.815	0.5214	0.03	2.84	0.23
	1.315		XXS		0.358	0.599	0.2817	0.01	3.66	0.12
1½	1.900			5S	0.065	1.770	2.4593	0.13	1.27	1.07
	1.900			10S	0.109	1.682	2.2209	0.12	2.08	0.96
	1.900	40	Std	40S	0.145	1.610	2.0348	0.11	2.72	0.88
	1.900	80	XS	80S	0.200	1.500	1.7663	0.09	3.63	0.77
	1.900	160			0.281	1.338	1.4053	0.07	4.86	0.61
	1.900		XXS		0.400	1.100	0.9499	0.05	6.41	0.41
2	2.375			5S	0.065	2.245	3.9564	0.21	1.60	1.71
	2.375			10S	0.109	2.157	3.6523	0.19	2.64	1.58
	2.375	40	Std	40S	0.154	2.067	3.3539	0.17	3.65	1.45
	2.375	80	XS	80S	0.218	1.939	2.9514	0.15	5.02	1.28
	2.375	160			0.343	1.689	2.2394	0.12	7.44	0.97
	2.375		XXS		0.436	1.503	1.7733	0.09	9.03	0.77
2½	2.875			5S	0.083	2.709	5.7609	0.30	2.47	2.50
	2.875			10S	0.12	2.635	5.4504	0.28	3.53	2.36
	2.875	40	Std		0.203	2.469	4.7853	0.25	5.79	2.07

(Continued)

(Continued)

Nominal Pipe Size NPS	Outside Diameter in.	Schedule a	Schedule b	Schedule c	Wall Thickness in.	Inside Diameter in.	Flow Area in.²	Volume gal/ft	Pipe Weight lb/ft	Water Weight* lb/ft
3	2.875	80	XS		0.276	2.323	4.2361	0.22	7.66	1.84
	2.875	160			0.375	2.125	3.5448	0.18	10.01	1.54
	2.875		XXS		0.552	1.771	2.4621	0.13	13.69	1.07
	3.5			5S	0.083	3.334	8.7257	0.45	3.03	3.78
	3.5			10S	0.120	3.260	8.3427	0.43	4.33	3.62
	3.5	40	Std	40S	0.216	3.068	7.3889	0.38	7.58	3.20
	3.5	80	XS	80S	0.300	2.900	6.6019	0.34	10.25	2.86
	3.5	160			0.437	2.626	5.4133	0.28	14.30	2.35
	3.5		XXS		0.600	2.300	4.1527	0.22	18.58	1.80
4	4.5			5S	0.083	4.334	14.7451	0.77	3.92	6.39
	4.5			10S	0.120	4.260	14.2459	0.74	5.61	6.17
	4.5	40	Std	40S	0.237	4.026	12.7238	0.66	10.79	5.51
	4.5	80	XS	80S	0.337	3.826	11.4910	0.60	14.98	4.98
	4.5	120			0.437	3.626	10.3211	0.54	18.96	4.47
	4.5	160			0.531	3.438	9.2786	0.48	22.51	4.02
	4.5		XXS		0.674	3.152	7.7991	0.41	27.54	3.38

NPS	OD	Sched No.	Std/XS/XXS	S						
6	6.625			5S	0.109	6.407	32.2240	1.67	7.59	13.96
	6.625			10S	0.134	6.357	31.7230	1.65	9.29	13.75
	6.625	40	Std	40S	0.280	6.065	28.8756	1.50	18.97	12.51
	6.625	80	XS	80S	0.432	5.761	26.0535	1.35	28.57	11.29
	6.625	120			0.562	5.501	23.7549	1.23	36.39	10.29
	6.625	160			0.718	5.189	21.1367	1.10	45.30	9.16
	6.625		XXS		0.864	4.897	18.8248	0.98	53.16	8.16
8	8.625			5S	0.109	8.407	55.4820	2.88	9.91	24.04
	8.625			10S	0.148	8.329	54.4572	2.83	13.40	23.60
	8.625	20			0.250	8.125	51.8223	2.69	22.36	22.46
	8.625	30			0.277	8.071	51.1357	2.66	24.70	22.16
	8.625	40	Std	40S	0.322	7.981	50.0016	2.60	28.55	21.67
	8.625	60			0.406	7.813	47.9187	2.49	35.64	20.76
	8.625	80	XS	80S	0.500	7.625	45.6404	2.37	43.39	19.78
	8.625	100			0.593	7.439	43.4409	2.26	50.87	18.82
	8.625	120			0.718	7.189	40.5702	2.11	60.63	17.58
	8.625	140			0.812	7.001	38.4760	2.00	67.76	16.67
	8.625		XXS		0.875	6.875	37.1035	1.93	72.42	16.08

(Continued)

(Continued)

Nominal Pipe Size NPS	Outside Diameter in.	Schedule a	Schedule b	Schedule c	Wall Thickness in.	Inside Diameter in.	Flow Area in.^{2}	Volume gal/ft	Pipe Weight lb/ft	Water Weight* lb/ft
10	8.625	160			0.906	6.813	36.4373	1.89	74.69	15.79
	10.75			5S	0.134	10.482	86.2498	4.48	15.19	37.37
	10.75			10S	0.165	10.420	85.2325	4.43	18.65	36.93
	10.75	20			0.250	10.250	82.4741	4.28	28.04	35.74
	10.75				0.279	10.192	81.5433	4.24	31.20	35.34
	10.75	30			0.307	10.136	80.6497	4.19	34.24	34.95
	10.75	40	Std	40S	0.365	10.020	78.8143	4.09	40.48	34.15
	10.75	60	XS	80S	0.500	9.750	74.6241	3.88	54.74	32.34
	10.75	80			0.593	9.564	71.8040	3.73	64.33	31.12
	10.75	100			0.718	9.314	68.0992	3.54	76.93	29.51
	10.75	120			0.843	9.064	64.4925	3.35	89.20	27.95
	10.75	140			1.000	8.750	60.1016	3.12	104.13	26.04
	10.75	160			1.125	8.500	56.7163	2.95	115.64	24.58
12	12.75			5S	0.156	12.438	121.4425	6.31	20.98	52.63
	12.75			10S	0.180	12.390	120.5070	6.26	24.16	52.22
	12.75	20			0.250	12.250	117.7991	6.12	33.38	51.05

OD	Sched. No.			Wall	ID				
12.75	30			0.330	12.090	114.7420	5.96	43.77	49.72
12.75		Std	40S	0.375	12.000	113.0400	5.87	49.56	48.98
12.75	40			0.406	11.938	111.8749	5.81	53.52	48.48
12.75		XS	80S	0.500	11.750	108.3791	5.63	65.42	46.96
12.75	60			0.562	11.626	106.1036	5.51	73.15	45.98
12.75	80			0.687	11.376	101.5895	5.28	88.51	44.02
12.75	100			0.843	11.064	96.0935	4.99	107.20	41.64
12.75	120			1.000	10.750	90.7166	4.71	125.49	39.31
12.75	140			1.125	10.500	86.5463	4.50	139.67	37.50
12.75	160			1.312	10.126	80.4907	4.18	160.27	34.88
14.00			5S	0.156	13.688	147.0787	7.64	23.07	63.73
14.00			10S	0.188	13.624	145.7065	7.57	27.73	63.14
14.00	10			0.250	13.500	143.0663	7.43	36.71	62.00
14.00	20			0.312	13.376	140.4501	7.30	45.61	60.86
14.00	30	Std		0.375	13.250	137.8166	7.16	54.57	59.72
14.00	40			0.437	13.126	135.2491	7.03	63.30	58.61
14.00		XS		0.500	13.000	132.6650	6.89	72.09	57.49
14.00				0.562	12.876	130.1462	6.76	80.66	56.40

14

(Continued)

(Continued)

Nominal Pipe Size NPS	Outside Diameter in.	Schedule a	Schedule b	Schedule c	Wall Thickness in.	Inside Diameter in.	Flow Area in.^{2}	Volume gal/ft	Pipe Weight lb/ft	Water Weight* lb/ft
	14.00	60			0.593	12.814	128.8959	6.70	84.91	55.85
	14.00				0.625	12.750	127.6116	6.63	89.28	55.30
	14.00				0.687	12.626	125.1415	6.50	97.68	54.23
	14.00	80			0.750	12.500	122.6563	6.37	106.13	53.15
	14.00				0.875	12.250	117.7991	6.12	122.65	51.05
	14.00	100			0.937	12.126	115.4263	6.00	130.72	50.02
	14.00	120			1.093	11.814	109.5629	5.69	150.67	47.48
	14.00	140			1.250	11.500	103.8163	5.39	170.21	44.99
	14.00	160			1.406	11.188	98.2595	5.10	189.11	42.58
16	16.00			5S	0.165	15.670	192.7559	10.01	27.90	83.53
	16.00			10S	0.188	15.624	191.6259	9.95	31.75	83.04
	16.00	10			0.250	15.500	188.5963	9.80	42.05	81.73
	16.00	20			0.312	15.376	185.5908	9.64	52.27	80.42
	16.00	30	Std		0.375	15.250	182.5616	9.48	62.58	79.11
	16.00				0.437	15.126	179.6048	9.33	72.64	77.83
	16.00	40	XS		0.500	15.000	176.6250	9.18	82.77	76.54

	16.00			0.562	14.876	173.7169	9.02	92.66	75.28
	16.00			0.625	14.750	170.7866	8.87	102.63	74.01
	16.00	60		0.656	14.688	169.3538	8.80	107.50	73.39
	16.00			0.687	14.626	167.9271	8.72	112.35	72.77
	16.00			0.750	14.500	165.0463	8.57	122.15	71.52
	16.00	80		0.843	14.314	160.8391	8.36	136.46	69.70
	16.00			0.875	14.250	159.4041	8.28	141.34	69.08
	16.00	100		1.031	13.938	152.5003	7.92	164.82	66.08
	16.00	120		1.218	13.564	144.4259	7.50	192.29	62.58
	16.00	140		1.437	13.126	135.2491	7.03	223.50	58.61
	16.00	160		1.593	12.814	128.8959	6.70	245.11	55.85
18	18.00		5S	0.165	17.670	245.0997	12.73	31.43	106.21
	18.00		10S	0.188	17.624	243.8252	12.67	35.76	105.66
	18.00	10		0.250	17.500	240.4063	12.49	47.39	104.18
	18.00	20		0.312	17.376	237.0114	12.31	58.94	102.70
	18.00		Std	0.375	17.250	233.5866	12.13	70.59	101.22
	18.00	30		0.437	17.126	230.2404	11.96	81.97	99.77
	18.00		XS	0.500	17.000	226.8650	11.79	93.45	98.31

(Continued)

(Continued)

Nominal Pipe Size NPS	Outside Diameter in.	Schedule a	Schedule b	Schedule c	Wall Thickness in.	Inside Diameter in.	Flow Area in.²	Volume gal/ft	Pipe Weight lb/ft	Water Weight* lb/ft
	18.00	40			0.562	16.876	223.5675	11.61	104.67	96.88
	18.00				0.625	16.750	220.2416	11.44	115.98	95.44
	18.00	60			0.687	16.626	216.9927	11.27	127.03	94.03
	18.00				0.750	16.500	213.7163	11.10	138.17	92.61
	18.00	80			0.875	16.250	207.2891	10.77	160.03	89.83
	18.00				0.937	16.126	204.1376	10.60	170.75	88.46
	18.00	100			1.156	15.688	193.1990	10.04	207.96	83.72
	18.00	120			1.375	15.250	182.5616	9.48	244.14	79.11
	18.00	140			1.562	14.876	173.7169	9.02	274.22	75.28
	18.00	160			1.781	14.438	163.6378	8.50	308.50	70.91
20	20.00			5S	0.188	19.624	302.3046	15.70	39.78	131.00
	20.00			10S	0.218	19.564	300.4588	15.61	46.06	130.20
	20.00	10			0.250	19.500	298.4963	15.51	52.73	129.35
	20.00				0.312	19.376	294.7121	15.31	65.60	127.71
	20.00	20	Std		0.375	19.250	290.8916	15.11	78.60	126.05
	20.00				0.437	19.126	287.1560	14.92	91.30	124.43

(Continued)

20.00	30	XS	0.500	19.000	283.3850	14.72	104.13	122.80
20.00			0.562	18.876	279.6982	14.53	116.67	121.20
20.00	40		0.593	18.814	277.8638	14.43	122.91	120.41
20.00			0.625	18.750	275.9766	14.34	129.33	119.59
20.00			0.687	18.626	272.3384	14.15	141.70	118.01
20.00			0.750	18.500	268.6663	13.96	154.19	116.42
20.00	60		0.812	18.376	265.0767	13.77	166.40	114.87
20.00			0.875	18.250	261.4541	13.58	178.72	113.30
20.00	80		1.031	17.938	252.5909	13.12	208.87	109.46
20.00	100		1.281	17.438	238.7058	12.40	256.10	103.44
20.00	120		1.500	17.000	226.8650	11.79	296.37	98.31
20.00	140		1.750	16.500	213.7163	11.10	341.09	92.61
20.00	160		1.968	16.064	202.5709	10.52	379.00	87.78
22.00		5S	0.188	21.624	367.0639	19.07	43.80	159.06
22.00		10S	0.218	21.564	365.0298	18.96	50.71	158.18
22.00	10		0.250	21.500	362.8663	18.85	58.07	157.24
22.00	20	Std	0.375	21.250	354.4766	18.41	86.61	153.61
22.00	30	XS	0.500	21.000	346.1850	17.98	114.81	150.01

22

(Continued)

Nominal Pipe Size NPS	Outside Diameter in.	Schedule a	Schedule b	Schedule c	Wall Thickness in.	Inside Diameter in.	Flow Area in.²	Volume gal/ft	Pipe Weight lb/ft	Water Weight* lb/ft
	22.00				0.625	20.750	337.9916	17.56	142.68	146.46
	22.00				0.750	20.500	329.8963	17.14	170.21	142.96
	22.00	60			0.875	20.250	321.8991	16.72	197.41	139.49
	22.00	80			1.125	19.750	306.1991	15.91	250.81	132.69
	22.00	100			1.375	19.250	290.8916	15.11	302.88	126.05
	22.00	120			1.625	18.750	275.9766	14.34	353.61	119.59
	22.00	140			1.875	18.250	261.4541	13.58	403.00	113.30
	22.00	160			2.125	17.750	247.3241	12.85	451.06	107.17
24	24.00			5S	0.188	23.624	438.1033	22.76	47.81	189.84
	24.00	10		10S	0.218	23.564	435.8807	22.64	55.37	188.88
	24.00				0.250	23.500	433.5163	22.52	63.41	187.86
	24.00	20			0.312	23.376	428.9533	22.28	78.93	185.88
	24.00		Std		0.375	23.250	424.3416	22.04	94.62	183.88
	24.00				0.437	23.126	419.8273	21.81	109.97	181.93
	24.00	30	XS		0.500	23.000	415.2650	21.57	125.49	179.95
	24.00				0.562	22.876	410.7994	21.34	140.68	178.01

	24.00	40	0.593	22.814	408.5757	21.22	148.24	177.05
	24.00	60	0.625	22.750	406.2866	21.11	156.03	176.06
	24.00	80	0.812	22.376	393.0380	20.42	201.09	170.32
	24.00	100	1.031	21.938	377.8015	19.63	252.91	163.71
	24.00	120	1.281	21.438	360.7765	18.74	310.82	156.34
	24.00	140	1.500	21.000	346.1850	17.98	360.45	150.01
	24.00	160	1.750	20.500	329.8963	17.14	415.85	142.96
	24.00		1.968	20.064	316.0128	16.42	463.07	136.94
26	26.00	10	0.250	25.500	510.4463	26.52	68.75	221.19
	26.00	20	0.312	25.376	505.4940	26.26	85.60	219.05
	26.00	Std	0.375	25.250	500.4866	26.00	102.63	216.88
	26.00	XS	0.500	25.000	490.6250	25.49	136.17	212.60
	26.00		0.625	24.750	480.8616	24.98	169.38	208.37
	26.00		0.750	24.500	471.1963	24.48	202.25	204.19
	26.00		0.875	24.250	461.6291	23.98	234.79	200.04
	26.00		1.000	24.000	452.1600	23.49	267.00	195.94
	26.00		1.125	23.750	442.7891	23.00	298.87	191.88
28	28.00		0.250	27.500	593.6563	30.84	74.09	257.25

(Continued)

(Continued)

Nominal Pipe Size NPS	Outside Diameter in.	Schedule a	Schedule b	Schedule c	Wall Thickness in.	Inside Diameter in.	Flow Area in.²	Volume gal/ft	Pipe Weight lb/ft	Water Weight* lb/ft
	28.00	10			0.312	27.376	588.3146	30.56	92.26	254.94
	28.00		Std		0.375	27.250	582.9116	30.28	110.64	252.60
	28.00	20	XS		0.500	27.000	572.2650	29.73	146.85	247.98
	28.00	30			0.625	26.750	561.7166	29.18	182.73	243.41
	28.00				0.750	26.500	551.2663	28.64	218.27	238.88
	28.00				0.875	26.250	540.9141	28.10	253.48	234.40
	28.00				1.000	26.000	530.6600	27.57	288.36	229.95
	28.00				1.125	25.750	520.5041	27.04	322.90	225.55
30	30.00			5S	0.250	29.500	683.1463	35.49	79.43	296.03
	30.00	10		10S	0.312	29.376	677.4153	35.19	98.93	293.55
	30.00		Std		0.375	29.250	671.6166	34.89	118.65	291.03
	30.00	20	XS		0.500	29.000	660.1850	34.30	157.53	286.08
	30.00	30			0.625	28.750	648.8516	33.71	196.08	281.17
	30.00	40			0.750	28.500	637.6163	33.12	234.29	276.30
	30.00				0.875	28.250	626.4791	32.54	272.17	271.47
	30.00				1.000	28.000	615.4400	31.97	309.72	266.69

NPS	OD	Schedule					
	30.00		1.125	27.750	604.4991	31.40	346.93	261.95
32	32.00		0.250	31.500	778.9163	40.46	84.77	337.53
	32.00	10	0.312	31.376	772.7959	40.15	105.59	334.88
	32.00	Std	0.375	31.250	766.6016	39.82	126.66	332.19
	32.00	XS	0.500	31.000	754.3850	39.19	168.21	326.90
	32.00	20	0.625	30.750	742.2666	38.56	209.43	321.65
	32.00	30	0.688	30.624	736.1961	38.24	230.08	319.02
	32.00	40	0.750	30.500	730.2463	37.93	250.31	316.44
	32.00		0.875	30.250	718.3241	37.32	290.86	311.27
	32.00		1.000	30.000	706.5000	36.70	331.08	306.15
	32.00		1.125	29.750	694.7741	36.09	370.96	301.07
34	34.00		0.250	33.500	880.9663	45.76	90.11	381.75
	34.00	10	0.312	33.376	874.4565	45.43	112.25	378.93
	34.00	Std	0.375	33.250	867.8666	45.08	134.67	376.08
	34.00	XS	0.500	33.000	854.8650	44.41	178.89	370.44
	34.00	20	0.625	32.750	841.9616	43.74	222.78	364.85
	34.00	30	0.688	32.624	835.4954	43.40	244.77	362.05
	34.00	40	0.750	32.500	829.1563	43.07	266.33	359.30

(Continued)

(Continued)

Nominal Pipe Size NPS	Outside Diameter in.	Schedule a	Schedule b	Schedule c	Wall Thickness in.	Inside Diameter in.	Flow Area in.^{2}	Volume gal/ft	Pipe Weight lb/ft	Water Weight* lb/ft
	34.00				0.875	32.250	816.4491	42.41	309.55	353.79
	34.00				1.000	32.000	803.8400	41.76	352.44	348.33
	34.00				1.125	31.750	791.3291	41.11	394.99	342.91
36	36.00				0.250	35.500	989.2963	51.39	95.45	428.70
	36.00	10			0.312	35.376	982.3972	51.03	118.92	425.71
	36.00		Std		0.375	35.250	975.4116	50.67	142.68	422.68
	36.00	20	XS		0.500	35.000	961.6250	49.95	189.57	416.70
	36.00	30			0.625	34.750	947.9366	49.24	236.13	410.77
	36.00	40			0.750	34.500	934.3463	48.54	282.35	404.88
	36.00				0.875	34.250	920.8541	47.84	328.24	399.04
	36.00				1.000	34.000	907.4600	47.14	373.80	393.23
	36.00				1.125	33.750	894.1641	46.45	419.02	387.47
42	42.00				0.250	41.500	1351.9663	70.23	111.47	585.85
	42.00		Std		0.375	41.250	1335.7266	69.39	166.71	578.81
	42.00	20	XS		0.500	41.000	1319.5850	68.55	221.61	571.82
	42.00	30			0.625	40.750	1303.5416	67.72	276.18	564.87

42.00	40	0.750	40.500	1287.5963	66.89	330.41	557.96
42.00		1.000	40.000	1256.0000	65.25	437.88	544.27
42.00		1.250	39.500	1224.7963	63.63	544.01	530.75
42.00		1.500	39.000	1193.9850	62.03	648.81	517.39
48	48	0.375	47.250	1753.4546	91.09	190.74	759.83
48		0.500	47.000	1734.9486	90.13	253.65	751.81
48		0.625	46.750	1716.5408	89.17	316.23	743.83
48		0.750	46.500	1698.2312	88.22	378.47	735.90
48		1.000	46.000	1661.9064	86.33	501.96	720.16
48		1.250	45.500	1625.9744	84.47	624.11	704.59
48		1.500	45.000	1590.4350	82.62	744.93	689.19
48		2.000	44.000	1520.5344	78.99	982.56	658.90

*Based on density of water of 62.4 lb/ft^3

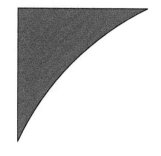

Appendix F

DOI: 10.1016/B978-1-85617-828-0.00017-2

Properties of Circular Pipes—SI Units

Nominal Pipe Size NPS	Outside Diameter in.	Outside Diameter mm	Wall Thickness mm	Inside Diameter mm	Flow Area m²	Volume m³/m	Pipe Weight kg/m	Water Weight* kg/m
½	0.84	21.34	1.651	18.034	0.0003	0.0003	0.80	0.26
	0.84	21.34	2.108	17.120	0.0002	0.0002	1.00	0.23
	0.84	21.34	2.769	15.799	0.0002	0.0002	1.27	0.20
	0.84	21.34	3.734	13.868	0.0002	0.0002	1.62	0.15
	0.84	21.34	4.750	11.836	0.0001	0.0001	1.94	0.11
	0.84	21.34	7.468	6.401	0.0000	0.0000	2.55	0.03
¾	1.05	26.67	1.651	23.368	0.0004	0.0004	1.02	0.43
	1.05	26.67	2.108	22.454	0.0004	0.0004	1.28	0.40
	1.05	26.67	2.870	20.930	0.0003	0.0003	1.68	0.34
	1.05	26.67	3.912	18.847	0.0003	0.0003	2.19	0.28
	1.05	26.67	5.537	15.596	0.0002	0.0002	2.88	0.19
	1.05	26.67	7.823	11.024	0.0001	0.0001	3.63	0.10
1	1.315	33.40	1.651	30.099	0.0007	0.0007	1.29	0.71
	1.315	33.40	2.769	27.864	0.0006	0.0006	2.09	0.61
	1.315	33.40	8.382	16.637	0.0002	0.0002	5.17	0.22
	1.315	33.40	4.547	24.308	0.0005	0.0005	3.23	0.46

1½	1.315	33.40	6.350	20.701	0.0003	0.0003	4.23	0.34
	1.315	33.40	9.093	15.215	0.0002	0.0002	5.44	0.18
	1.9	48.26	1.651	44.958	0.0016	0.0016	1.90	1.59
	1.9	48.26	2.769	42.723	0.0014	0.0014	3.10	1.43
	1.9	48.26	3.683	40.894	0.0013	0.0013	4.04	1.31
	1.9	48.26	5.080	38.100	0.0011	0.0011	5.40	1.14
	1.9	48.26	7.137	33.985	0.0009	0.0009	7.23	0.91
	1.9	48.26	10.160	27.940	0.0006	0.0006	9.53	0.61
2	2.375	60.33	1.651	57.023	0.0026	0.0026	2.39	2.55
	2.375	60.33	2.769	54.788	0.0024	0.0024	3.92	2.36
	2.375	60.33	3.912	52.502	0.0022	0.0022	5.44	2.16
	2.375	60.33	5.537	49.251	0.0019	0.0019	7.47	1.91
	2.375	60.33	8.712	42.901	0.0014	0.0014	11.08	1.45
	2.375	60.33	11.074	38.176	0.0011	0.0011	13.43	1.14
2½	2.875	73.03	2.108	68.809	0.0037	0.0037	3.68	3.72
	2.875	73.03	3.048	66.929	0.0035	0.0035	5.25	3.52
	2.875	73.03	5.156	62.713	0.0031	0.0031	8.62	3.09
	2.875	73.03	7.010	59.004	0.0027	0.0027	11.40	2.73
	2.875	73.03	9.525	53.975	0.0023	0.0023	14.90	2.29
	2.875	73.03	14.021	44.983	0.0016	0.0016	20.38	1.59

(Continued)

(Continued)

Nominal Pipe Size NPS	Outside Diameter in.	Outside Diameter mm	Wall Thickness mm	Inside Diameter mm	Flow Area m²	Volume m³/m	Pipe Weight kg/m	Water Weight* kg/m
3	3.5	88.90	2.108	84.684	0.0056	0.0056	4.51	5.63
	3.5	88.90	3.048	82.804	0.0054	0.0054	6.45	5.39
	3.5	88.90	5.486	77.927	0.0048	0.0048	11.27	4.77
	3.5	88.90	7.620	73.660	0.0043	0.0043	15.25	4.26
	3.5	88.90	11.100	66.700	0.0035	0.0035	21.27	3.49
	3.5	88.90	15.240	58.420	0.0027	0.0027	27.65	2.68
4	4.5	114.30	2.108	110.084	0.0095	0.0095	5.83	9.52
	4.5	114.30	3.048	108.204	0.0092	0.0092	8.35	9.20
	4.5	114.30	6.020	102.260	0.0082	0.0082	16.05	8.21
	4.5	114.30	8.560	97.180	0.0074	0.0074	22.29	7.42
	4.5	114.30	11.100	92.100	0.0067	0.0067	28.21	6.66
	4.5	114.30	13.487	87.325	0.0060	0.0060	33.49	5.99
	4.5	114.30	17.120	80.061	0.0050	0.0050	40.98	5.03
6	6.625	168.28	2.769	162.738	0.0208	0.0208	11.29	20.80
	6.625	168.28	3.404	161.468	0.0205	0.0205	13.82	20.48
	6.625	168.28	7.112	154.051	0.0186	0.0186	28.23	18.64
	6.625	168.28	10.973	146.329	0.0168	0.0168	42.51	16.82

8	6.625	168.28	14.275	139.725	0.0153	0.0153	54.14	15.33
	6.625	168.28	18.237	131.801	0.0136	0.0136	67.39	13.64
	6.625	168.28	21.946	124.384	0.0122	0.0122	79.09	12.15
	8.625	219.08	2.769	213.538	0.0358	0.0358	14.75	35.81
	8.625	219.08	3.759	211.557	0.0352	0.0352	19.94	35.15
	8.625	219.08	6.350	206.375	0.0335	0.0335	33.27	33.45
	8.625	219.08	7.036	205.003	0.0330	0.0330	36.74	33.01
	8.625	219.08	8.179	202.717	0.0323	0.0323	42.48	32.28
	8.625	219.08	10.312	198.450	0.0309	0.0309	53.02	30.93
	8.625	219.08	12.700	193.675	0.0295	0.0295	64.55	29.46
	8.625	219.08	15.062	188.951	0.0280	0.0280	75.69	28.04
	8.625	219.08	18.237	182.601	0.0262	0.0262	90.21	26.19
	8.625	219.08	20.625	177.825	0.0248	0.0248	100.81	24.84
	8.625	219.08	22.225	174.625	0.0239	0.0239	107.76	23.95
	8.625	219.08	23.012	173.050	0.0235	0.0235	111.13	23.52
10	10.75	273.05	3.404	266.243	0.0557	0.0557	22.60	55.67
	10.75	273.05	4.191	264.668	0.0550	0.0550	27.75	55.02
	10.75	273.05	6.350	260.350	0.0532	0.0532	41.71	53.24
	10.75	273.05	7.087	258.877	0.0526	0.0526	46.42	52.64
	10.75	273.05	7.798	257.454	0.0521	0.0521	50.94	52.06

(Continued)

(Continued)

Nominal Pipe Size NPS	Outside Diameter in.	Outside Diameter mm	Wall Thickness mm	Inside Diameter mm	Flow Area m²	Volume m³/m	Pipe Weight kg/m	Water Weight* kg/m
	10.75	273.05	9.271	254.508	0.0509	0.0509	60.23	50.87
	10.75	273.05	12.700	247.650	0.0482	0.0482	81.44	48.17
	10.75	273.05	15.062	242.926	0.0463	0.0463	95.71	46.35
	10.75	273.05	18.237	236.576	0.0440	0.0440	114.46	43.96
	10.75	273.05	21.412	230.226	0.0416	0.0416	132.71	41.63
	10.75	273.05	25.400	222.250	0.0388	0.0388	154.93	38.79
	10.75	273.05	28.575	215.900	0.0366	0.0366	172.06	36.61
12	12.75	323.85	3.962	315.925	0.0784	0.0784	31.22	78.39
	12.75	323.85	4.572	314.706	0.0778	0.0778	35.95	77.79
	12.75	323.85	6.350	311.150	0.0760	0.0760	49.66	76.04
	12.75	323.85	8.382	307.086	0.0741	0.0741	65.13	74.06
	12.75	323.85	9.525	304.800	0.0730	0.0730	73.74	72.97
	12.75	323.85	10.312	303.225	0.0722	0.0722	79.64	72.21
	12.75	323.85	12.700	298.450	0.0700	0.0700	97.33	69.96
	12.75	323.85	14.275	295.300	0.0685	0.0685	108.84	68.49
	12.75	323.85	17.450	288.950	0.0656	0.0656	131.69	65.57

(Continued)

	12.75	323.85	21.412	281.026	0.0620	0.0620	159.50	62.03
	12.75	323.85	25.400	273.050	0.0586	0.0586	186.71	58.56
	12.75	323.85	28.575	266.700	0.0559	0.0559	207.82	55.86
	12.75	323.85	33.325	257.200	0.0520	0.0520	238.46	51.96
	14	355.60	3.962	347.675	0.0949	0.0949	34.32	94.94
	14	355.60	4.775	346.050	0.0941	0.0941	41.26	94.05
	14	355.60	6.350	342.900	0.0923	0.0923	54.62	92.35
	14	355.60	7.925	339.750	0.0907	0.0907	67.86	90.66
	14	355.60	9.525	336.550	0.0890	0.0890	81.19	88.96
	14	355.60	11.100	333.400	0.0873	0.0873	94.18	87.30
	14	355.60	12.700	330.200	0.0856	0.0856	107.26	85.63
	14	355.60	14.275	327.050	0.0840	0.0840	120.01	84.01
	14	355.60	15.062	325.476	0.0832	0.0832	126.33	83.20
	14	355.60	15.875	323.850	0.0824	0.0824	132.83	82.37
	14	355.60	17.450	320.700	0.0808	0.0808	145.33	80.78
	14	355.60	19.050	317.500	0.0792	0.0792	157.91	79.17
	14	355.60	22.225	311.150	0.0760	0.0760	182.49	76.04
	14	355.60	23.800	308.000	0.0745	0.0745	194.50	74.51
	14	355.60	27.762	300.076	0.0707	0.0707	224.17	70.72
14	14	355.60	31.750	292.100	0.0670	0.0670	253.25	67.01

(Continued)

Nominal Pipe Size NPS	Outside Diameter in.	Outside Diameter mm	Wall Thickness mm	Inside Diameter mm	Flow Area m²	Volume m³/m	Pipe Weight kg/m	Water Weight* kg/m
16	14	355.60	35.712	284.175	0.0634	0.0634	281.37	63.43
	16	406.40	4.191	398.018	0.1244	0.1244	41.52	124.42
	16	406.40	4.775	396.850	0.1237	0.1237	47.24	123.69
	16	406.40	6.350	393.700	0.1217	0.1217	62.57	121.74
	16	406.40	7.925	390.550	0.1198	0.1198	77.78	119.80
	16	406.40	9.525	387.350	0.1178	0.1178	93.11	117.84
	16	406.40	11.100	384.200	0.1159	0.1159	108.07	115.93
	16	406.40	12.700	381.000	0.1140	0.1140	123.15	114.01
	16	406.40	14.275	377.850	0.1121	0.1121	137.87	112.13
	16	406.40	15.875	374.650	0.1102	0.1102	152.70	110.24
	16	406.40	16.662	373.075	0.1093	0.1093	159.95	109.32
	16	406.40	17.450	371.500	0.1084	0.1084	167.17	108.40
	16	406.40	19.050	368.300	0.1065	0.1065	181.75	106.54
	16	406.40	21.412	363.576	0.1038	0.1038	203.04	103.82
	16	406.40	22.225	361.950	0.1029	0.1029	210.30	102.89
	16	406.40	26.187	354.025	0.0984	0.0984	245.24	98.44

(Continued)

16	406.40	30.937	344.526	0.0932	0.0932	286.10	93.23
16	406.40	36.500	333.400	0.0873	0.0873	332.54	87.30
16	406.40	40.462	325.476	0.0832	0.0832	364.69	83.20
18	457.20	4.191	448.818	0.1582	0.1582	46.76	158.21
18	457.20	4.775	447.650	0.1574	0.1574	53.21	157.39
18	457.20	6.350	444.500	0.1552	0.1552	70.51	155.18
18	457.20	7.925	441.350	0.1530	0.1530	87.69	152.99
18	457.20	9.525	438.150	0.1508	0.1508	105.02	150.78
18	457.20	11.100	435.000	0.1486	0.1486	121.96	148.62
18	457.20	12.700	431.800	0.1464	0.1464	139.04	146.44
18	457.20	14.275	428.650	0.1443	0.1443	155.73	144.31
18	457.20	15.875	425.450	0.1422	0.1422	172.56	142.16
18	457.20	17.450	422.300	0.1401	0.1401	189.00	140.07
18	457.20	19.050	419.100	0.1380	0.1380	205.58	137.95
18	457.20	22.225	412.750	0.1338	0.1338	238.11	133.80
18	457.20	23.800	409.600	0.1318	0.1318	254.05	131.77
18	457.20	29.362	398.475	0.1247	0.1247	309.41	124.71
18	457.20	34.925	387.350	0.1178	0.1178	363.24	117.84
18	457.20	39.675	377.850	0.1121	0.1121	408.00	112.13
18	457.20	45.237	366.725	0.1056	0.1056	459.01	105.63

18

(Continued)

Nominal Pipe Size NPS	Outside Diameter in.	Outside Diameter mm	Wall Thickness mm	Inside Diameter mm	Flow Area m²	Volume m³/m	Pipe Weight kg/m	Water Weight* kg/m
20	20	508.00	4.775	498.450	0.1951	0.1951	59.19	195.13
	20	508.00	5.537	496.926	0.1939	0.1939	68.53	193.94
	20	508.00	6.350	495.300	0.1927	0.1927	78.46	192.68
	20	508.00	7.925	492.150	0.1902	0.1902	97.61	190.23
	20	508.00	9.525	488.950	0.1878	0.1878	116.94	187.77
	20	508.00	11.100	485.800	0.1854	0.1854	135.85	185.36
	20	508.00	12.700	482.600	0.1829	0.1829	154.93	182.92
	20	508.00	14.275	479.450	0.1805	0.1805	173.59	180.54
	20	508.00	15.062	477.876	0.1794	0.1794	182.87	179.36
	20	508.00	15.875	476.250	0.1781	0.1781	192.42	178.14
	20	508.00	17.450	473.100	0.1758	0.1758	210.83	175.79
	20	508.00	19.050	469.900	0.1734	0.1734	229.42	173.42
	20	508.00	20.625	466.750	0.1711	0.1711	247.58	171.10
	20	508.00	22.225	463.550	0.1688	0.1688	265.91	168.77
	20	508.00	26.187	455.625	0.1630	0.1630	310.77	163.04
	20	508.00	32.537	442.925	0.1541	0.1541	381.03	154.08

	20	508.00	38.100	431.800	0.1464	0.1464	440.96	146.44
22	20	508.00	44.450	419.100	0.1380	0.1380	507.50	137.95
	20	508.00	49.987	408.026	0.1308	0.1308	563.90	130.76
	22	558.80	4.775	549.250	0.2369	0.2369	65.16	236.94
	22	558.80	5.537	547.726	0.2356	0.2356	75.45	235.62
	22	558.80	6.350	546.100	0.2342	0.2342	86.40	234.23
	22	558.80	9.525	539.750	0.2288	0.2288	128.86	228.81
	22	558.80	12.700	533.400	0.2235	0.2235	170.82	223.46
	22	558.80	15.875	527.050	0.2182	0.2182	212.28	218.17
	22	558.80	19.050	520.700	0.2129	0.2129	253.25	212.94
	22	558.80	22.225	514.350	0.2078	0.2078	293.72	207.78
	22	558.80	28.575	501.650	0.1976	0.1976	373.17	197.65
	22	558.80	34.925	488.950	0.1878	0.1878	450.64	187.77
	22	558.80	41.275	476.250	0.1781	0.1781	526.12	178.14
	22	558.80	47.625	463.550	0.1688	0.1688	599.61	168.77
	22	558.80	53.975	450.850	0.1596	0.1596	671.12	159.64
24	24	609.60	4.775	600.050	0.2828	0.2828	71.14	282.79
	24	609.60	5.537	598.526	0.2814	0.2814	82.38	281.36
	24	609.60	6.350	596.900	0.2798	0.2798	94.35	279.83
	24	609.60	7.925	593.750	0.2769	0.2769	117.44	276.88

(Continued)

(Continued)

Nominal Pipe Size NPS	Outside Diameter in.	Outside Diameter mm	Wall Thickness mm	Inside Diameter mm	Flow Area m²	Volume m³/m	Pipe Weight kg/m	Water Weight* kg/m
	24	609.60	9.525	590.550	0.2739	0.2739	140.78	273.91
	24	609.60	11.100	587.400	0.2710	0.2710	163.62	270.99
	24	609.60	12.700	584.200	0.2680	0.2680	186.71	268.05
	24	609.60	14.275	581.050	0.2652	0.2652	209.31	265.17
	24	609.60	15.062	579.476	0.2637	0.2637	220.56	263.73
	24	609.60	15.875	577.850	0.2623	0.2623	232.15	262.25
	24	609.60	20.625	568.350	0.2537	0.2537	299.19	253.70
	24	609.60	26.187	557.225	0.2439	0.2439	376.30	243.87
	24	609.60	32.537	544.525	0.2329	0.2329	462.46	232.88
	24	609.60	38.100	533.400	0.2235	0.2235	536.30	223.46
	24	609.60	44.450	520.700	0.2129	0.2129	618.73	212.94
	24	609.60	49.987	509.626	0.2040	0.2040	688.99	203.98
26	26	660.40	6.350	647.700	0.3295	0.3295	102.29	329.49
	26	660.40	7.925	644.550	0.3263	0.3263	127.36	326.29
	26	660.40	9.525	641.350	0.3231	0.3231	152.70	323.06
	26	660.40	12.700	635.000	0.3167	0.3167	202.60	316.69

28	26	660.40	15.875	628.650	0.3104	0.3104	252.01	310.39
	26	660.40	19.050	622.300	0.3042	0.3042	300.92	304.15
	26	660.40	22.225	615.950	0.2980	0.2980	349.34	297.98
	26	660.40	25.400	609.600	0.2919	0.2919	397.26	291.86
	26	660.40	28.575	603.250	0.2858	0.2858	444.68	285.82
	28	711.20	6.350	698.500	0.3832	0.3832	110.24	383.20
	28	711.20	7.925	695.350	0.3798	0.3798	137.27	379.75
	28	711.20	9.525	692.150	0.3763	0.3763	164.61	376.26
	28	711.20	12.700	685.800	0.3694	0.3694	218.49	369.39
	28	711.20	15.875	679.450	0.3626	0.3626	271.87	362.58
	28	711.20	19.050	673.100	0.3558	0.3558	324.76	355.84
	28	711.20	22.225	666.750	0.3492	0.3492	377.15	349.15
	28	711.20	25.400	660.400	0.3425	0.3425	429.04	342.54
	28	711.20	28.575	654.050	0.3360	0.3360	480.43	335.98
30	30	762.00	6.350	749.300	0.4410	0.4410	118.18	440.96
	30	762.00	7.925	746.150	0.4373	0.4373	147.19	437.26
	30	762.00	9.525	742.950	0.4335	0.4335	176.53	433.52
	30	762.00	12.700	736.600	0.4261	0.4261	234.38	426.14
	30	762.00	15.875	730.250	0.4188	0.4188	291.74	418.83
	30	762.00	19.050	723.900	0.4116	0.4116	348.59	411.57

(Continued)

(Continued)

Nominal Pipe Size NPS	Outside Diameter in.	Outside Diameter mm	Wall Thickness mm	Inside Diameter mm	Flow Area m²	Volume m³/m	Pipe Weight kg/m	Water Weight* kg/m
	30	762.00	22.225	717.550	0.4044	0.4044	404.95	404.39
	30	762.00	25.400	711.200	0.3973	0.3973	460.82	397.26
	30	762.00	28.575	704.850	0.3902	0.3902	516.19	390.20
32	32	812.80	6.350	800.100	0.5028	0.5028	126.13	502.78
	32	812.80	7.925	796.950	0.4988	0.4988	157.10	498.83
	32	812.80	9.525	793.750	0.4948	0.4948	188.45	494.83
	32	812.80	12.700	787.400	0.4869	0.4869	250.27	486.95
	32	812.80	15.875	781.050	0.4791	0.4791	311.60	479.12
	32	812.80	17.475	777.850	0.4752	0.4752	342.32	475.21
	32	812.80	19.050	774.700	0.4714	0.4714	372.43	471.37
	32	812.80	22.225	768.350	0.4637	0.4637	432.76	463.67
	32	812.80	25.400	762.000	0.4560	0.4560	492.60	456.04
	32	812.80	28.575	755.650	0.4485	0.4485	551.94	448.47
34	34	863.60	6.350	850.900	0.5687	0.5687	134.07	568.65
	34	863.60	7.925	847.750	0.5645	0.5645	167.02	564.45
	34	863.60	9.525	844.550	0.5602	0.5602	200.37	560.20

36	34	863.60	12.700	838.200	0.5518	0.5518	266.16	551.81
	34	863.60	15.875	831.850	0.5435	0.5435	331.46	543.48
	34	863.60	17.475	828.650	0.5393	0.5393	364.18	539.30
	34	863.60	19.050	825.500	0.5352	0.5352	396.26	535.21
	34	863.60	22.225	819.150	0.5270	0.5270	460.57	527.01
	34	863.60	25.400	812.800	0.5189	0.5189	524.38	518.87
	34	863.60	28.575	806.450	0.5108	0.5108	587.69	510.79
	36	914.40	6.350	901.700	0.6386	0.6386	142.02	638.58
	36	914.40	7.925	898.550	0.6341	0.6341	176.93	634.13
	36	914.40	9.525	895.350	0.6296	0.6296	212.28	629.62
	36	914.40	12.700	889.000	0.6207	0.6207	282.05	620.72
	36	914.40	15.875	882.650	0.6119	0.6119	351.32	611.88
	36	914.40	19.050	876.300	0.6031	0.6031	420.10	603.11
	36	914.40	22.225	869.950	0.5944	0.5944	488.38	594.40
	36	914.40	25.400	863.600	0.5858	0.5858	556.16	585.76
	36	914.40	28.575	857.250	0.5772	0.5772	623.45	577.17
42	42	1066.80	6.350	1054.100	0.8727	0.8727	165.85	872.68
	42	1066.80	9.525	1047.750	0.8622	0.8622	248.04	862.20
	42	1066.80	12.700	1041.400	0.8518	0.8518	329.72	851.78
	42	1066.80	15.875	1035.050	0.8414	0.8414	410.91	841.42

(Continued)

(Continued)

Nominal Pipe Size NPS	Outside Diameter in.	Outside Diameter mm	Wall Thickness mm	Inside Diameter mm	Flow Area m²	Volume m³/m	Pipe Weight kg/m	Water Weight* kg/m
	42	1066.80	19.050	1028.700	0.8311	0.8311	491.61	831.13
	42	1066.80	25.400	1016.000	0.8107	0.8107	651.50	810.73
	42	1066.80	31.750	1003.300	0.7906	0.7906	809.41	790.59
	42	1066.80	38.100	990.600	0.7707	0.7707	965.34	770.70
48	48	1219.20	9.525	1200.150	1.1313	1.1313	283.79	1131.26
	48	1219.20	12.700	1193.800	1.1193	1.1193	377.39	1119.32
	48	1219.20	15.875	1187.450	1.1074	1.1074	470.50	1107.44
	48	1219.20	19.050	1181.100	1.0956	1.0956	563.11	1095.63
	48	1219.20	25.400	1168.400	1.0722	1.0722	746.84	1072.20
	48	1219.20	31.750	1155.700	1.0490	1.0490	928.59	1049.01
	48	1219.20	38.100	1143.000	1.0261	1.0261	1108.35	1026.09
	48	1219.20	50.800	1117.600	0.9810	0.9810	1461.91	980.99

*Based on density of water of 1000 kg/m³.

Appendix G

DOI: 10.1016/B978-1-85617-828-0.00018-4

Head Loss in Water Pipes—USCS Units

Nominal Pipe Size NPS	Outside Diameter in.	Wall Thickness in.	Flow Rate gpm	Velocity ft/s	Head Loss[*] ft/1000 ft
½	0.840	0.109	15	15.84	2245.37
1	1.315	0.330	15	14.28	1745.63
1½	1.900	0.145	75	11.82	430.82
2	2.375	0.154	150	14.34	460.62
2½	2.875	0.203	200	13.40	330.26
3	3.500	0.216	250	10.85	173.35
3½	4.000	0.226	300	9.74	119.71
4	4.500	0.237	400	10.08	110.21
6	6.625	0.280	500	5.55	22.65
8	8.625	0.322	1000	6.41	21.47
10	10.75	0.365	1500	6.10	15.03
12	12.75	0.250	3000	8.17	20.39
	12.75	0.312	4000	11.11	36.50
	12.75	0.344	4500	12.63	46.58
	12.75	0.375	5000	14.18	58.05
	12.75	0.406	5000	14.33	59.53
	12.75	0.500	5000	14.79	64.32
14	14.00	0.250	6000	13.45	45.85
	14.00	0.312	6000	13.70	47.96
	14.00	0.375	6000	13.96	50.22
	14.00	0.500	6000	14.50	55.10
16	16.00	0.250	8000	13.60	39.86
	16.00	0.281	8000	13.71	40.65
	16.00	0.312	8000	13.82	41.45
	16.00	0.375	8000	14.05	43.14
	16.00	0.500	8000	14.52	46.76

(Continued)

Nominal Pipe Size NPS	Outside Diameter in.	Wall Thickness in.	Flow Rate gpm	Velocity ft/s	Head Loss* ft/1000 ft
18	18.00	0.250	10000	13.34	33.37
	18.00	0.281	10000	13.43	33.95
	18.00	0.312	10000	13.53	34.54
	18.00	0.500	10000	14.13	38.43
	18.00	0.625	10000	14.56	41.30
20	20.00	0.250	12000	12.89	27.61
	20.00	0.375	12000	13.23	29.40
	20.00	0.500	12000	13.58	31.34
	20.00	0.625	12000	13.94	33.42
	20.00	0.812	12000	14.52	36.87
22	22.00	0.250	15000	13.26	25.95
	22.00	0.375	15000	13.57	27.47
	22.00	0.500	15000	13.89	29.10
	22.00	0.625	15000	14.23	30.84
	22.00	0.750	15000	14.58	32.72
24	24.00	0.218	20000	14.71	28.29
	24.00	0.312	20000	14.95	29.41
	24.00	0.500	20000	15.44	31.83
	24.00	0.562	20000	15.61	32.68
	24.00	0.625	20000	15.79	33.57
26	26.00	0.312	20000	12.69	19.72
	26.00	0.500	20000	13.07	21.21
	26.00	0.625	20000	13.34	22.27
	26.00	0.750	20000	13.61	23.40
	26.00	0.875	20000	13.89	24.60
28	28.00	0.312	25000	13.63	20.60

(Continued)

(Continued)

Nominal Pipe Size NPS	Outside Diameter in.	Wall Thickness in.	Flow Rate gpm	Velocity ft/s	Head Loss* ft/1000 ft
	28.00	0.500	25000	14.01	22.04
	28.00	0.625	25000	14.27	23.06
	28.00	0.750	25000	14.54	24.14
	28.00	0.875	25000	14.82	25.28
30	30.00	0.312	30000	14.20	20.48
	30.00	0.500	30000	14.57	21.81
	30.00	0.625	30000	14.83	22.75
	30.00	0.750	30000	15.09	23.74
	30.00	0.875	30000	15.36	24.78
32	32.00	0.312	30000	12.45	14.86
	32.00	0.500	30000	12.75	15.76
	32.00	0.625	30000	12.96	16.40
	32.00	0.750	30000	13.17	17.06
	32.00	0.875	30000	13.39	17.76
34	34.00	0.312	35000	12.83	14.64
	34.00	0.500	35000	13.13	15.47
	34.00	0.625	35000	13.33	16.05
	34.00	0.750	35000	13.54	16.66
	34.00	0.875	35000	13.75	17.30
36	36.00	0.312	40000	13.06	14.12
	36.00	0.500	40000	13.34	14.87
	36.00	0.625	40000	13.53	15.40
	36.00	0.750	40000	13.73	15.95
	36.00	0.875	40000	13.93	16.53
42	42.00	0.375	50000	12.00	10.10
	42.00	0.500	50000	12.15	10.40
	42.00	0.625	50000	12.30	10.72

(Continued)

Nominal Pipe Size NPS	Outside Diameter in.	Wall Thickness in.	Flow Rate gpm	Velocity ft/s	Head Loss* ft/1000 ft
	42.00	0.750	50000	12.45	11.04
	42.00	1.000	55000	14.04	14.00
48	48.00	0.375	60000	10.98	7.31
	48.00	0.500	60000	11.10	7.50
	48.00	0.625	60000	11.21	7.70
	48.00	0.750	60000	11.34	7.90
	48.00	1.000	60000	11.58	8.33

*Head Loss: Based on the Hazen-Williams formula with C = 120.
For other values of C, multiply head loss by $(120/C)^{1.852}$.

Head Loss in Water Pipes—SI Units

Nominal Pipe Size NPS	Outside Diameter mm	Wall Thickness mm	Flow Rate m^3/h	Velocity m/s	Head Loss* m/km
½	21.34	1.651	4	4.35	1587.44
1	33.40	1.651	8	3.12	473.55
1½	48.26	5.080	16	3.90	542.31
2	60.33	5.537	32	4.66	560.51
2½	73.03	5.156	50	4.50	394.92
3	88.9	5.486	75	4.37	290.66
4	114.3	8.560	100	3.75	168.97
6	168.28	10.973	200	3.30	83.10
8	219.08	10.312	400	3.59	68.03
10	273.05	15.062	600	3.60	53.85
12	323.85	3.962	1000	3.54	38.58
	323.85	6.350	1000	3.65	41.55
	323.85	9.525	1000	3.81	45.94

(Continued)

(Continued)

Nominal Pipe Size NPS	Outside Diameter mm	Wall Thickness mm	Flow Rate m³/h	Velocity m/s	Head Loss* m/km
	323.85	12.700	1000	3.97	50.90
	323.85	17.450	1000	4.24	59.58
	323.85	25.400	1000	4.74	78.49
14	355.60	3.962	1200	3.51	33.92
	355.60	9.525	1200	3.75	39.74
	355.60	15.062	1200	4.01	46.77
	355.60	31.750	1200	4.97	79.22
16	406.40	4.775	1500	3.37	26.92
	406.40	9.525	1500	3.54	30.30
	406.40	14.275	1500	3.72	34.19
	406.40	17.450	1500	3.84	37.13
	406.40	36.500	1500	4.77	62.89
18	457.20	6.350	1800	3.22	21.72
	457.20	12.700	1800	3.41	25.02
	457.20	19.050	1800	3.62	28.93
	457.20	23.800	1800	3.79	32.35
	457.20	39.675	1800	4.46	47.92
20	508.00	6.350	2200	3.17	18.60
	508.00	12.700	2200	3.34	21.11
	508.00	19.050	2200	3.52	24.03
	508.00	38.100	2200	4.17	36.28
	508.00	49.987	2200	4.67	47.80
22	558.80	4.775	2600	3.05	15.32
	558.80	6.350	2600	3.08	15.75
	558.80	12.700	2600	3.23	17.66
	558.80	22.225	2600	3.48	21.09
	558.80	47.625	2600	4.28	34.99

(Continued)

Nominal Pipe Size NPS	Outside Diameter mm	Wall Thickness mm	Flow Rate m^3/h	Velocity m/s	Head Loss* m/km
24	609.60	4.775	3200	3.14	14.62
	609.60	6.350	3200	3.18	15.00
	609.60	12.700	3200	3.32	16.66
	609.60	26.187	3200	3.64	20.97
	609.60	49.987	3200	4.36	32.40
26	660.40	6.350	3800	3.20	13.86
	660.40	9.525	3800	3.27	14.54
	660.40	15.875	3800	3.40	16.03
	660.40	22.225	3800	3.54	17.70
	660.40	28.575	3800	3.69	19.59
28	711.20	6.350	4500	3.26	13.12
	711.20	9.525	4500	3.32	13.72
	711.20	15.875	4500	3.45	15.01
	711.20	22.225	4500	3.58	16.46
	711.20	28.575	4500	3.72	18.07
30	762.00	6.350	6000	3.78	15.88
	762.00	9.525	6000	3.84	16.55
	762.00	15.875	6000	3.98	18.00
	762.00	22.225	6000	4.12	19.61
	762.00	28.575	6000	4.27	21.39
32	812.80	6.350	7500	4.14	17.44
	812.80	9.525	7500	4.21	18.13
	812.80	15.875	7500	4.35	19.61
	812.80	22.225	7500	4.49	21.25
	812.80	28.575	7500	4.65	23.04
34	863.60	6.350	9000	4.40	18.12
	863.60	9.525	9000	4.46	18.79

(Continued)

(Continued)

Nominal Pipe Size NPS	Outside Diameter mm	Wall Thickness mm	Flow Rate m³/h	Velocity m/s	Head Loss* m/km
	863.60	15.875	9000	4.60	20.23
	863.60	22.225	9000	4.74	21.80
	863.60	28.575	9000	4.89	23.53
36	914.40	6.350	10200	4.44	17.22
	914.40	9.525	10200	4.50	17.83
	914.40	15.875	10200	4.63	19.11
	914.40	22.225	10200	4.77	20.51
	914.40	28.575	10200	4.91	22.03
42	1066.80	6.350	12000	3.82	10.88
	1066.80	9.525	12000	3.87	11.20
	1066.80	15.875	12000	3.96	11.89
	1066.80	25.400	12000	4.11	13.01
	1066.80	38.100	12000	4.33	14.72
48	1219.20	9.525	15000	3.68	8.74
	1219.20	15.875	15000	3.76	9.21
	1219.20	25.400	15000	3.89	9.96
	1219.20	38.100	15000	4.06	11.09
	1219.20	50.800	15000	4.25	12.37

*Head Loss: Based on the Hazen-Williams formula with C = 120
For other values of C, multiply head loss by $(120/C)^{1.852}$

Appendix H

DOI: 10.1016/B978-1-85617-828-0.00019-6

Darcy Friction Factors*

Reynolds Number	Friction Factor		
R	e/D = 0.0001	e/D = 0.0002	e/D = 0.0003
5000	0.0380	0.0602	0.0725
10,000	0.0311	0.0559	0.0703
20,000	0.0261	0.0524	0.0684
30,000	0.0237	0.0506	0.0675
40,000	0.0222	0.0495	0.0669
50,000	0.0212	0.0487	0.0664
60,000	0.0204	0.0481	0.0661
70,000	0.0198	0.0476	0.0658
80,000	0.0192	0.0471	0.0656
90,000	0.0188	0.0468	0.0654
100,000	0.0185	0.0465	0.0652
125,000	0.0177	0.0458	0.0648
150,000	0.0172	0.0454	0.0646
200,000	0.0164	0.0447	0.0642
225,000	0.0161	0.0444	0.0640
250,000	0.0158	0.0442	0.0639
275,000	0.0156	0.0440	0.0638
300,000	0.0154	0.0438	0.0637
325,000	0.0153	0.0436	0.0636
350,000	0.0151	0.0435	0.0635
375,000	0.0150	0.0434	0.0634
400,000	0.0149	0.0433	0.0634
425,000	0.0147	0.0432	0.0633
450,000	0.0146	0.0431	0.0632
500,000	0.0145	0.0429	0.0631
750,000	0.0139	0.0423	0.0628
1,000,000	0.0135	0.0419	0.0626

*Friction factor based on the Swamee-Jain equation: $f = 0.25/[Log10(e/3.7D + 5.74/R^{0.9})]^2$.

Appendix I

Least Squares Method

In Chapter 2, the least squares method (LSM) was used to model the pump head versus capacity (H-Q) curve and the efficiency versus capacity (E-Q) curve as a second-degree polynomial with the following format:

$$H = a_0 + a_1Q + a_2Q^2 \qquad \text{from Equation (2.1)}$$

$$E = b_0 + b_1Q + b_2Q^2 \qquad \text{from Equation (2.2)}$$

where a_0, a_1, and a_2 are constants for the H-Q equation, and b_0, b_1, and b_2 are constants for the E-Q equation.

© 2010 E. Shashi Menon. Published by Elsevier Inc. All right reserved.
DOI: 10.1016/B978-1-85617-828-0.00020-2

From a given set of data for H and E versus Q, a second-degree curve fit can be performed using LSM to obtain the constants a_0, a_1, a_2 and b_0, b_1, and b_2, as described next. Suppose there are n sets of H, Q data and n sets of E, Q data. The following values are calculated from a given data:

Σx_i = sum of all Q values from i = 1 to n

Σy_i = sum of all H values from i = 1 to n

$\Sigma x_i y_i$ = sum of all (Q × H) values from i = 1 to n

$\Sigma x_i^2 y_i$ = sum of all (Q^2 × H) values from i = 1 to n

Σx_i^3 = sum of all Q^3 values from i = 1 to n

Σx_i^4 = sum of all Q^4 values from i = 1 to n

These sums are then used in three simultaneous equations to solve for the constants a_0, a_1, and a_2 as follows:

$$a_0 + a_1(\Sigma x_i) + a_2(\Sigma x^2_i) = \Sigma y_i \tag{I.1}$$

$$a_0(\Sigma x_i) + a_1(\Sigma x^2_i) + a_2(\Sigma x^3_i) = \Sigma x_i y_i \tag{I.2}$$

$$a_0(\Sigma x^2_i) + a_1(\Sigma x^3_i) + a_2(\Sigma x^4_i) = \Sigma x^2_i y_i \tag{I.3}$$

Solving for the constants a_0, a_1, and a_2 will produce the best parabolic curve fit for the H-Q curve.

Similarly, for the E-Q curve, the procedure is repeated for the sums replacing H values with E values for y. This results in the three simultaneous equations in b_0, b_1, and b_2 as follows:

$$b_0 + b_1(\Sigma x_i) + b_2(\Sigma x^2_i) = \Sigma y_i \tag{I.4}$$

$$b_0(\Sigma x_i) + b_1(\Sigma x^2_i) + b_2(\Sigma x^3_i) = \Sigma x_i y_i \tag{I.5}$$

$$b_0(\Sigma x^2_i) + b_1(\Sigma x^3_i) + b_2(\Sigma x^4_i) = \Sigma x^2_i y_i \tag{I.6}$$

$$a_0 = \text{Det}_0/\text{Det}_C \quad a_1 = -\text{Det}_1/\text{Det}_C \quad a_2 = \text{Det}_2/\text{Det}_C$$

where Det_0, Det_1, Det_2, and Det_C are the determinants for Equations (I.1) through (I.3), as follows:

$$Det_0 = \begin{vmatrix} \sum x_i & \sum x_i^2 & \sum y_i \\ \sum x_i^2 & \sum x_i^3 & \sum x_i y_i \\ \sum x_i^3 & \sum x_i^4 & \sum x_i^2 y_i \end{vmatrix}$$

$$Det_1 = \begin{vmatrix} 1 & \sum x_i^2 & \sum y_i \\ \sum x_i & \sum x_i^3 & \sum x_i y_i \\ \sum x_i^2 & \sum x_i^4 & \sum x_i^2 y_i \end{vmatrix}$$

$$Det_2 = \begin{vmatrix} 1 & \sum x_i & \sum y_i \\ \sum x_i & \sum x_i^2 & \sum x_i y_i \\ \sum x_i^2 & \sum x_i^3 & \sum x_i^2 y_i \end{vmatrix}$$

$$Det_C = \begin{vmatrix} 1 & \sum x_i & \sum x_i^2 \\ \sum x_i & \sum x_i^2 & \sum x_i^3 \\ \sum x_i^2 & \sum x_i^3 & \sum x_i^4 \end{vmatrix}$$

Similarly,

$$b_0 = Det_0/Det_C \qquad b_1 = -Det_1/Det_C \qquad b_2 = Det_2/Det_C$$

where Det_0, Det_1, Det_2, and Det_C are the determinants for Equations (I.4) through (I.6), as follows:

$$Det_0 = \begin{vmatrix} \sum x_i & \sum x_i^2 & \sum y_i \\ \sum x_i^2 & \sum x_i^3 & \sum x_i y_i \\ \sum x_i^3 & \sum x_i^4 & \sum x_i^2 y_i \end{vmatrix}$$

$$Det_1 = \begin{vmatrix} 1 & \sum x_i^2 & \sum y_i \\ \sum x_i & \sum x_i^3 & \sum x_i y_i \\ \sum x_i^2 & \sum x_i^4 & \sum x_i^2 y_i \end{vmatrix}$$

$$
\mathrm{Det}_2 = \begin{vmatrix} 1 & \sum x_i & \sum y_i \\ \sum x_i & \sum x_i^2 & \sum x_i y_i \\ \sum x_i^2 & \sum x_i^3 & \sum x_i^2 y_i \end{vmatrix}
$$

$$
\mathrm{Det}_C = \begin{vmatrix} 1 & \sum x_i & \sum x_i^2 \\ \sum x_i & \sum x_i^2 & \sum x_i^3 \\ \sum x_i^2 & \sum x_i^3 & \sum x_i^4 \end{vmatrix}
$$

The LSM may also be used in conjunction with an Excel spreadsheet that can be downloaded from the publisher's website for simulating this and other calculations in the book.

References

1. Garr M. Jones, et al. *Pumping Station Design*, Revised Third Edition. Elsevier, Inc., 2008.
2. Tyler G. Hicks, et al. *Pump Application Engineering*. McGraw-Hill, 1971.
3. V. S. Lobanoff and R. R. Ross. *Centrifugal Pumps Design & Application*. Gulf Publishing, 1985.
4. E. F. Brater and H. W. King. *Handbook of Hydraulics*. McGraw-Hill, 1982.
5. Alfred Benaroya. *Fundamentals and Application of Centrifugal Pumps*. The Petroleum Publishing Company, 1978.
6. *Flow of Fluids through Valves, Fittings and Pipes*. Crane Company, 1976.
7. *Hydraulic Institute Engineering Data Book*. Hydraulic Institute, 1979.
8. Igor J. Karassik, et al. *Pump Handbook*. McGraw-Hill, 1976.
9. Michael Volk. *Pump Characteristics and Applications*, Second Edition. CRC Press, Taylor and Francis Group, 2005.

Index

Printed and bound by CPI Group (UK) Ltd, Croydon, CR0 4YY

03/10/2024

01040435-0005